MATHEMATICS

WORKBOOK

for

WELDING

by

J.L. McCabe

Names of persons, places or things used through-
out this book are fictitious in nature and are
not intended to refer to specific places or things
or persons living or dead. Any similarity is
entirely coincidental.

Third Edition
1990

Printed in the United States of America

EVERYDAY BOOKS
811 Moundridge Dr.
Lawrence, Kansas 66049
913/841-7643

ISBN 0-942465-05-9

FOREWORD

"Mathematics for Welding" is a comprehensive self-study of the application of mathematics for those entering the field of welding.

Developing computation skills in the many forms of mathematics presented in this book are most important to the success of the welder. The welding fabricator will find it necessary to solve the problem before beginning the fabrication process.

After research and careful planning, this book was written for students learning welding fabrication and for those already on the job and interested in advancement. This book is intended for use during training and afterwards as a reference guide. Keep it in good order.

From the earlier units in basic math through the later units in the application of basic algebra and geometry, the importance in developing these skills becomes more apparent as application problems are presented.

The material in this book is presented in the order in which your "hands-on" training is offered. More complex welding techniques and processes are learned as you progress through the program. This book was written to keep pace with that part of your training.

Material presented in this book is intended to help you reach a higher level of competency so that you may have a better opportunity to reach your goals.

Take pride in your work...

Aurora, Colorado   J.L. McCabe
1987

CONTENTS

UNIT 1

WHOLE NUMBERS

ADDITION:

1.   19
     21
   + 5

2.   39
     17
   +10

3.   321
      14
   +110

4.   791
      47
   + 23

5.   1007
      941
    + 39

6.   47
      910
        7
     +213

7.   3710
       90
       17
    + 410

8.   36190
       412
        71
     + 9152

9.   27959
      1007
       312
     +   19

10.   17
        412
       1007
      + 409

SUBTRACTION:

1. 25
   -19

2. 317
   - 98

3. 417
   - 19

4. 312
   - 79

5. 9351
   - 429

6. 712
   - 19

7. 9171
   - 312

8. 5170
   - 99

9. 31942
   - 919

10. 73541
    - 9179

11. 4009
    -   10

12. 49170
    -   705

MULTIPLICATION:

1. 96
   x10

2. 371
   x 43

3. 75
   x 9

4. 217
   x 36

5. 191
   x740

6. 512
   x 12

7. 317
   x412

8. 4712
   x 319

9. 3997
   x 497

10. 21701
    x 912

11. 309
    x 27

12. 9007
    x 315

DIVISION:

1. $9\overline{)36}$　　　2. $7\overline{)56}$　　　3. $11\overline{)33}$　　　4. $30\overline{)90}$

5. $3\overline{)333}$　　　6. $4\overline{)450}$　　　7. $7\overline{)350}$　　　8. $14\overline{)540}$

9. $25\overline{)750}$　　　10. $12\overline{)600}$

MIXED OPERATIONS:

1. 17 + 43 - 10 =　　　　　　2. 37 - 19 + 15 =

3. 35 + 15 - 10 =　　　　　　4. 15 x 5 ÷ 3 =

5. 18 ÷ 3 x 6 =　　　　　　6. 20 x 5 ÷ 5 =

7. 18 + 4 - 6 x 2 ÷ 4 =　　　　8. 27 - 7 + 5 x 3 ÷ 25 =

9. 35 - 15 + 10 ÷ 3 + 25 =　　　10. 412 ÷ 4 x 7 + 2 - 3 =

FRACTIONS

When <u>adding</u> and <u>subtracting</u> fractions, the denominator <u>must</u> be the same. When necessary, change to least common denominator as shown in problem #1 and always simplify your answer.

Addition:

1. $\dfrac{1}{16} = \dfrac{1}{16}$

$+\dfrac{5}{8} = \dfrac{10}{16}$

$\dfrac{11}{16}$

2. $\dfrac{1}{16}$

$+\dfrac{9}{16}$

3. $\dfrac{3}{32}$

$+\dfrac{5}{32}$

4. $\dfrac{11}{64}$

$+\dfrac{7}{64}$

5. $\dfrac{1}{4}$

$+\dfrac{3}{4}$

6. $\dfrac{21}{64}$

$+\dfrac{7}{64}$

7. $\dfrac{3}{16}$

$+\dfrac{5}{32}$

8. $\dfrac{7}{8}$

$+\dfrac{5}{16}$

9. $\dfrac{3}{32}$

$\dfrac{5}{16}$

$+\dfrac{9}{64}$

10. $3\dfrac{1}{2}$

$7\dfrac{3}{16}$

$+\dfrac{9}{32}$

11. $5\dfrac{1}{4}$

$2\dfrac{7}{16}$

$+3\dfrac{1}{2}$

12. $7\dfrac{5}{64}$

$3\dfrac{3}{8}$

$+5\dfrac{9}{16}$

13. $19\dfrac{3}{4}$

$+\dfrac{5}{8}$

14. $\dfrac{7}{64}$

$+1\dfrac{13}{32}$

15. $9\dfrac{3}{8}$

$+5\dfrac{5}{16}$

4

Subtraction: When necessary change to lease common denominator
<u>before</u> proceeding, as in problem #1.

1. $\dfrac{7}{8} = \dfrac{14}{16}$ 2. $\dfrac{3}{4}$ 3. $\dfrac{9}{16}$ 4. $\dfrac{11}{32}$

$-\dfrac{7}{16}\ \dfrac{7}{16}$ $-\dfrac{1}{8}$ $-\dfrac{3}{8}$ $-\dfrac{3}{32}$

$\dfrac{7}{16}$

5. $\dfrac{17}{32}$ 6. $3\dfrac{1}{2}$ 7. $2\dfrac{7}{8}$ 8. $\dfrac{19}{64}$

$-\dfrac{3}{64}$ $-1\dfrac{7}{8}$ $-1\dfrac{15}{16}$ $-\dfrac{7}{32}$

9. $5\dfrac{7}{8}$ 10. $3\dfrac{1}{2}$ 11. $15\dfrac{7}{8}$ 12. $10\dfrac{9}{16}$

$-\dfrac{15}{16}$ $-1\dfrac{3}{4}$ $-11\dfrac{9}{16}$ $-9\dfrac{13}{16}$

13. $4\dfrac{1}{2}$ 14. $17\dfrac{9}{64}$ 15. $4\dfrac{1}{8}$ 16. $7\dfrac{1}{8}$

$-1\dfrac{1}{2}$ $-11\dfrac{7}{32}$ $-3\dfrac{7}{8}$ -5

5

Multiplication: Denominators need not be the same here.
Just multiply straight across as in problem #1. Mixed
numbers must be changed to improper fractions before
proceeding as in problem #7. The answer here is then
changed back to a mixed number.

1. $\frac{3}{8} \times \frac{1}{8} = \frac{3}{64}$

2. $\frac{3}{32} \times \frac{1}{2} =$

3. $\frac{1}{16} \times \frac{3}{8} =$

4. $\frac{5}{8} \times \frac{7}{16} =$

5. $\frac{1}{2} \times \frac{5}{16} =$

6. $\frac{3}{8} \times \frac{1}{32} =$

7. $4\frac{1}{2} \times 1\frac{1}{8} = \frac{9}{2} \times \frac{9}{8} = \frac{81}{16} = 5\frac{1}{16}$

8. $7\frac{3}{16} \times 1\frac{1}{8} =$

9. $3\frac{5}{16} \times 5 =$

10. $9 \times \frac{7}{16} =$

11. $9\frac{7}{32} \times 4\frac{1}{2} \times \frac{1}{2} =$

12. $3 \times 7\frac{3}{16} \times 2 =$

13. $5\frac{1}{16} \times \frac{9}{32} \times 7 =$

14. $3\frac{1}{8} \times 7\frac{5}{32} \times 2 =$

6

Division: Invert the <u>second fraction and multiply</u> straight across. Simplify the answer as in problem #1. Mixed numbers <u>must</u> be changed to improper fractions <u>before</u> proceeding.

1. $\frac{7}{8} \div \frac{1}{2} = \frac{7}{8} \times \frac{2}{1} = \frac{14}{8} = 1\frac{6}{8} = 1\frac{3}{4}$

2. $\frac{3}{16} \div \frac{1}{4} =$

3. $\frac{9}{32} \div \frac{3}{64} =$

4. $\frac{7}{16} \div 3 =$

5. $\frac{31}{32} \div \frac{1}{4} =$

6. $1\frac{1}{2} \div \frac{1}{4} =$

7. $\frac{7}{8} \div 3\frac{1}{2} =$

8. $7\frac{3}{16} \div 3\frac{1}{2} =$

9. $4\frac{1}{4} \div 2\frac{3}{8} =$

10. $3\frac{3}{4} \div 6\frac{1}{4} =$

7

Mixed operations: Work these problems as shown in problem #1.
If necessary, ask for help from your teacher to get started.

1. $1\frac{3}{4} + \frac{1}{2} - \frac{3}{8} = 1\frac{3}{4} = 1\frac{3}{4}$ $1\frac{5}{4} = \frac{10}{8}$

$+\frac{1}{2} = \frac{2}{4}$ $-\frac{3}{8} = \frac{3}{8}$

$1\frac{5}{4} - \frac{3}{8} = \quad 1 \qquad \frac{7}{8}$ answer

2. $3\frac{3}{4} - \frac{1}{2} \times 1\frac{1}{8} \div \frac{3}{4} + 3\frac{1}{2} =$

3. $\frac{7}{8} + 1\frac{1}{2} - \frac{3}{4} \times 1\frac{1}{8} \div \frac{1}{2} =$

4. $\frac{5}{32} + \frac{1}{16} - \frac{3}{8} \times \frac{5}{16} \div \frac{1}{8} =$

5. $6\frac{1}{2} - 3 \times \frac{1}{8} \div \frac{1}{16} + 2\frac{1}{4} \times 2 =$

6. $\frac{3}{4} - \frac{1}{8} + 2 \div \frac{1}{2} \times \frac{7}{16} =$

UNIT 3

DECIMAL FRACTIONS

Addition: (decimal points must be in line)

1. .37	2. .095	3. 1.750	4. 71.5	5. 11.6
+.09	+ .7	+ .09	+ .92	+ 7.

6. 417.0	7. 37.501	8. 6	9. 926.50
+ .91	+ 7	+50.10	+ 2.5

10.   7.9
    +1.2

Subtraction: (decimal points must be in line)

1. .96	2. 1.75	3. 25	4. .17	5. 37.0
- .7	- .60	-9.1	-5.1	- 19

6. 712	7. 31.9	8. 747.09	9. 475.60
-515.90	- 9.0	- 9.7	- 109

10.   417.9
    -    7.3

9

Multiplication: (count over from right to left to position the decimal point in the answer).

1.   3.7   2.   .50   3.   39.0   4.   46   5.   312
   x 9   x1.7   x 7.7   x.9   x.66

6.   517   7.   5.199   8.   7.52   9.   695.0   10.   1.97
   x.92   x   .99   x.312   x .97   x .27

Division:

1. $.79 \div 6 =$ 2. $9.72 \div 16 =$

3. $3.50 \div 5 =$ 4. $4.12 \div 17 =$

5. $31.8 \div .6 =$ 6. $.160 \div 5 =$

7. $959.9 \div .14 =$ 8. $.65 \div 5. =$

9. $13.0 \div 5 =$ 10. $35. \div .50 =$

Mixed Operations:

1. 9.795 + .059 - 1.3 =

2. .175 - .07 + 3.17 x 9.5 =

3. 3.49 + 1.75 - .175 x 3.5 ÷ 1.3 =

4. 71.25 x .375 + 5.25 - 3.10 ÷ 2.5 =

5. 109.9 + .171 x 3 ÷ 150.5 - 7.5 =

6. 75.5 x 3.25 - 15.5 + 7.5 ÷ 3.15 =

Unit 4

Fractions and Decimal Conversions

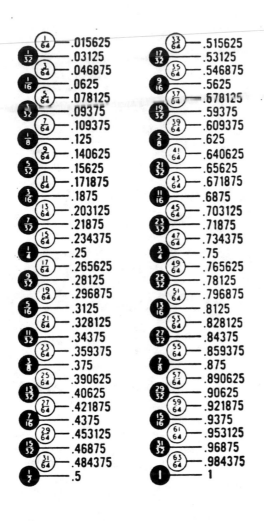

CONVERSION CHART

Use for <u>checking</u> your answers.
The numbers in this chart represent values less than one (1).

1. To convert a fraction to its decimal equivalent, follow this procedure.

2. Divide the numerator by the denominator.

3. Example:

$\frac{1}{4}$ numerator
denominator

$$4\overline{\smash{)}1.00} \quad .25 \text{ or } .250 \text{ or } .2500$$
$$\frac{8}{20}$$
$$\frac{20}{}$$

4. The answers all have the same value.

5. When setting up the division problem a decimal point must be placed to the right of the dividend because it is a whole number.

6. Example:

divisor ⟶ 4 |1.00 .25 quotient
dividend
place decimal   8
point here   20
                  20

7. The quotient (answer to a division problem) is read as follows:

   .25 is read 25 hundredths
   .250 is read 250 thousandths
   .2500 is read 2 thousand 5 hundred ten thousandths

   These values are all equal.

8. To convert a <u>mixed number</u> to its decimal equivalent follow this procedure.

9. Example: A mixed number consists of a whole number and a <u>proper</u> fraction

   whole number → 1 $\frac{1}{4}$ ← proper fraction

10. Convert the mixed number to an <u>improper</u> fraction and set up the division problem.

11. 1-1/4 = $\frac{5}{4}$ multiply denominator (4) by the whole number (1) and add the numerator (1), answer is $\frac{5}{4}$.

    This is called an <u>improper</u> fraction because the numerator is <u>larger</u> than the denominator.

12.
    $$\begin{array}{r} 1.25 \\ 4\overline{)5.00} \\ \underline{4} \\ 10 \\ \underline{8} \\ 20 \\ \underline{20} \end{array}$$
    this is read 1 and 25 hundredths

13. To convert a decimal to its fractional equivalent, follow this procedure.

14. Place the decimal number as a numerator (minus decimal point) over a denominator with the same number of zeros as the number of digits in the numerator and place the number 1 (one) to the left of the zeros.

    Example:

    $$.25 = \frac{25}{100}$$

13

14. (continued)

Now reduce this fraction to its __lowest__ form.

Divide by 5) $\frac{25}{100}$ = 5) $\frac{5}{20}$ = $\frac{1}{4}$ the fractional equivalent
of the decimal .25

Use __any__ number as a divisor that __does__ __not__ leave a remainder.

Therefore .25 = $\frac{1}{4}$

Example:

.625 = 5) $\frac{625}{1000}$ = 5) $\frac{125}{200}$ = 5) $\frac{25}{40}$ = $\frac{5}{8}$

 divide by

Therefore .625 = $\frac{5}{8}$

15. Convert to inches:

.9' (this reads $\frac{9}{10}$ of a foot)

Use ratio and proportion

9:10::x:12   set the ratio

$\frac{9}{10}$ = $\frac{x}{12}$ set the proportion

10x = 108   cross multiply

$\frac{10x}{10}$ = $\frac{108}{10}$ divide each side by 10

$$10 \overline{)108.} \quad 10.8$$
$$\underline{10}$$
$$8\ 0$$
$$\underline{8\ 0}$$

x = $\frac{108}{10}$

x = 10.8" or 10 $\frac{51}{64}$"   (See chart on page 12)

14

Convert to tenths of a foot:

8"

Use ratio and proportion

8:12::x:10   set the ratio

$\frac{8}{12} = \frac{x}{10}$ set the proportion

12x = 80   cross multiply

$\frac{12x}{12} = \frac{80}{12}$ divide both sides by 12

```
        6.66
    12 )80.
        72
         8 0
         7 2
           80
           72
```

```
                    .666
    6.66 ÷ 10 = 10 )6.660
                    6 0
                     66
                     60
                     60
```

x = .67' (rounded)

Write out the following decimal fractions in words:

1.   1.71

2.   .710

3.   3.04

4.   7.005

5.   90.07

Conversions:

Convert the following fractions to decimals:

1. $\dfrac{1}{16}$ =   2. $\dfrac{3}{16}$ =

3. $\dfrac{7}{16}$ =   4. $\dfrac{11}{16}$ =

5. $\dfrac{1}{8}$ =   6. $3\dfrac{3}{8}$ =

7. $\dfrac{7}{8}$ =   8. $\dfrac{11}{8}$ =

9. $\dfrac{1}{32}$ =   10. $1\dfrac{3}{32}$ =

11. $5\dfrac{7}{32}$ =   12. $\dfrac{11}{32}$ =

13. $\dfrac{1}{64}$ =   14. $\dfrac{3}{64}$ =

15. $\dfrac{7}{64}$ =   16. $1\dfrac{11}{64}$ =

Convert the following decimals to fractions:

1. .750 =   2. .625 =

3. .315 =   4. .875 =

5. .550 =   6. .250 =

7. .28125 =   8. .3125 =

9. .40625 =   10. .4375 =

11. .50 =   12. .9375 =

13. .8125 =   14. .6875 =

15. 1.375 =   16. 2.0625 =

Convert to inches:

    1. .9' = (this reads 9/10 of a foot)

    2. .7' =

    3. .8' =

    4. .4' =

    5. .3' =

    6. .5' =

Convert to tenths of a foot:

    1. 9" =

    2. 6" =

    3. 3" =

    4. 12" =

    5. 2" =

    6. 1'4" =

Another method to convert a decimal to a __specific__ fraction:

$$.792 = \frac{}{32}$$

Multiply .792 by 32 = 25.344 round to 25. The closest fraction in 32nds would be 25/32.

$$.519 = \frac{}{16}$$

.519 x 16 = 8.304 round to 8. The closest fraction in 16ths is 8/16 or 1/2.

$$.519 = \frac{}{32}$$

.519 x 32 = 16.608 round to 17. The closest fraction in 32nds is 17/32.

$$.519 = \frac{}{64}$$

.519 x 64 = 33.216 round to 33. The closest fraction in 64ths is 33/64.

$$.625 = \frac{}{8}$$

.625 x 8 = 5. The closest fraction in 8ths is 5/8.

Check your answers on the chart.

Convert the following decimal fractions:

1. .792 = $\dfrac{}{64}$

2. .597 = $\dfrac{}{16}$

3. .375 = $\dfrac{}{8}$

4. .625 = $\dfrac{}{16}$

5. .3125 = $\dfrac{}{16}$

6. .951 = $\dfrac{}{64}$

7. .775 = $\dfrac{}{16}$

8. .5125 = $\dfrac{}{32}$

9. .675 = $\dfrac{}{16}$

10. .292 = $\dfrac{}{16}$

To convert an improper fraction to a decimal:

$$1\frac{1}{16} = \frac{17}{16}$$

$$\begin{array}{r} 1.0625 \\ 16\overline{\smash{)}17.0000} \\ \underline{16} \\ 1\ 00 \\ \underline{96} \\ 40 \\ \underline{32} \\ 80 \end{array}$$

Now check .0625 on the chart to confirm that this is the equivalent of 1/16. Now add the whole number 1 to get 1-1/16.

Convert the following to decimal fractions: any mixed numbers should be changed to improper fractions before proceeding.

1. $3\frac{7}{32}$" =

2. $\frac{9}{16}$" =

3. $1\frac{7}{8}$" =

4. $5\frac{1}{2}$" =

5. $\frac{1}{16}$" =

6. $7\frac{5}{8}$" = .

7. $2\frac{3}{16}$" =

8. $15\frac{1}{16}$" =

21

1. Convert 27' to inches _____

2. Convert 65" to feet and inches _____

3. Convert 4'7" to total inches _____

4. A piece of I-beam is 9'7-1/2" long. Change the dimension to total inches. _____

5. Convert 98" to feet and inches _____

6. Change the following to feet, inches and fractions of an inch.

   a. 27.625" _____

   b. 37.125" _____

   c. 13.750" _____

   d. 19.375" _____

UNIT 5

Metrics

^{1}Kilo	Hecto	Deka	Meter	Deci	Centi	Milli
=1000	=100	=10	=1	$=\frac{1}{10}$	$=\frac{1}{100}$	$=\frac{1}{1000}$

Metric Value Line

Some conversion examples:

1m = 10 decimeters
1m = 100 centimeters
1m = 1000 millimters
1m = .1 dekameters
1m = .01 hectometers
1m = .001 kilometers

a. To convert 47m to mm, multiply by 1000 since 1m = 1000 mm, 47m = 47 (1000) mm = 47000mm

b. 247mm = 247 (.001) m = .247 m

c. Use the value line as follows:

To convert 47m to mm requires 3 decimal moves to the <u>right</u>, 47.000 = 47,000mm

To convert 47,000mm to kilometers requires 6 decimal moves to the <u>left</u>, 47,000mm = .047 km

5m = 500cm,   10cm = 100 mm

d. Therefore, to convert between units the change occurs through movement of the decimal point on the value line.

e. If the digit is a whole number as 9, place a decimal point to the <u>right</u>: such as 9. If this is 9 meters and should be expressed as kilometers do this: 9. = .009 (the decimals point was moved 3 places to the left).

f.   Therefore, by using the value line we find that:

.009km = .09hm, .9dm, 9m, 90dm, 900cm, 9000mm

All of these values were obtained by moving the decimal point to the <u>right</u>. <u>All</u> of the values are equal.

Then 900cm is the same as 9m; we moved the "inserted" decimal point 2 places to the <u>left</u> because meter is two places to the <u>left</u> of centimeter (cm) on the line.

g.   Therefore, 9.3m = 930 cm. The decimal point was moved 2 places to the <u>right</u>. The decimal point should be dropped if there are no digits to the right of the decimal point in the answer.

Convert the following:

10m = _____cm

25cm = _____mm

100mm = _____m

500km = _____m

700cm = _____km

5 yds. = _____m

10 miles = _____km

7m = _____inches

24

METRIC CONVERSION

Metric to English			English to Metric		
to convert from	to	multiply by	to convert from	to	multiply by
meters	yards	1.09	yards	meters	.914
meters	feet	3.28	feet	meters	.305
meters	inches	39.37	inches	meters	.0254
kilometers	miles	.62	miles	kilometers	1.609
grams	pounds	.0022	pounds	grams	.454
kilograms	pounds	2.20	pounds	kilograms	.454
liters	quarts	1.06	quarts	liters	.946
liters	gallons	.264	gallons	liters	3.79

CONVERTED VALUES

1 kilometer (km)	.621 miles
1 meter (m)	39.37 inches
1 centimeter (cm)	0.3937 inches
1 millimeter (mm)	0.03937 inches
25.4 mm	1.0000 inch
1 liter	1.06 quarts
3.785 liters	1 gallon
1 kilogram (kg)	2.2 pounds
0.454 kg	1 pound

MEASURES OF CAPACITY

1 liter (=1 cubic decimeter) = 61.023 cubic inch
.03531 cubic feet
.2642 gallon
2.202 lbs. of water at 62°

28.317 liters = 1 cubic foot
3.785 liters = 1 gallon

TEMPERATURE

To convert from Fahrenheit to Celsius

$$C = \frac{5(F-32)}{9}$$

To convert from Celsius to Fahrenheit

$$F = \frac{9}{5}C + 32$$

Find the missing dimensions in the drawing shown here.

a. A = 30cm, B = 18cm, D = 95cm

Find dimension c _____cm

b. B = 60 cm., C = 70cm, D = 200cm

Find dimension A _____cm

c. C = 25mm, B = 20mm, A = 30mm

Find dimension D _____mm

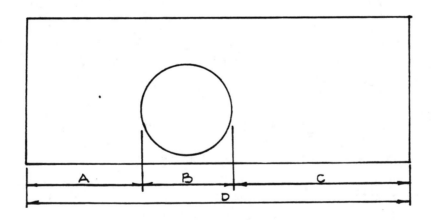

1. The steel tube shown here is 12 centimeters long. What is the length in millimeters?

   Solution:

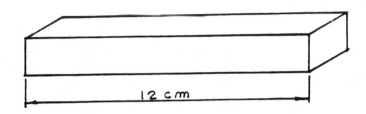

2. The angle iron shown is 5 meters long. What is the equivalent length in inches?

   Solution:

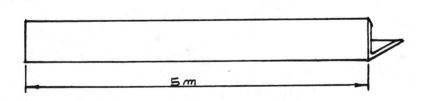

UNIT 6

APPLICATION PROBLEMS

1. A welding shop inventory includes the following: wide flange beam--97', I-beam--27', angle iron--139', channel iron--412'. Find the total number of feet in inventory.

   Solution:

2. Total the number of feet of angle iron in the following orders. 17'3", 19'9", 7'5", 14'7", 31'10" and 6'2".

   Solution:

Determine the distance from the start of this tape to letters
A, B, C, D, E, F, and reduce the fraction to its lowest form.
(Each division is 1/8").

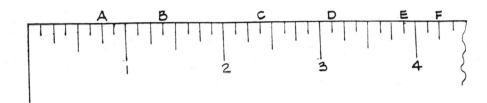

A _____

B _____

C _____

D _____

E _____

F _____

Using the steel tape in the preceding problem, determine
the distance __between__ the following letters.

A to B _____

A to C _____

B to C _____

B to D _____

A to D _____

C to E _____

C to D _____

D to E _____

D to F _____

C to F _____

E to F _____

1. Add   1'7"
                +13'7-5/8"

2. Subtract   13'7-1/2"
                - 6'9-3/4"

3. Add   6' 4-7/8"
                +9'10-1/8"

4. Subtract   7'3-1/2"
                -6'9-5/8"

5. Add   5'6"
                +1'9"

6. Subtract · 3'9"
                -0'7"

7. Add   1'11"
                +0' 9"

Tolerances:

Determine the longest and shortest length for the dimensions shown:

a. 19.750"   + .0375   longest _____
                - .0375   shortest _____

b. 7.40"   + .250   longest _____
                - .125   shortest _____

c. 17'8-1/4"   + 1/8   longest _____
                - 1/16   shortest _____

d. 3'7-1/2"   + .625   longest _____
                - .125   shortest _____

e. 5.970"   + .0125   longest _____
                - .750   shortest _____

1. Five pieces of angle iron have the following dimensions,
   7-1/2", 8-3/8", 1-7/8", 3-5/8" and 3-1/4". What is the
   total length in inches?

   Solution:

2. Five holes are drilled in the piece of plate shown. What
   is the total distance between holes?

   Solution:

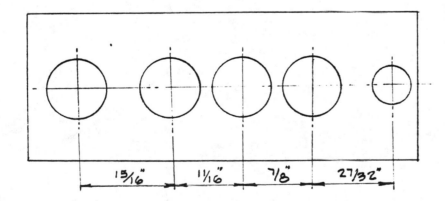

1. An angle iron has the five welded sections as shown. What is the total completed length?

   Solution:

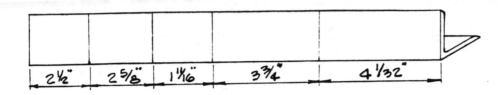

2. In the above problem, add 5/8" to each section shown and then determine the total completed length.

   Solution:

1. Find the missing dimension in the part shown.

    Solution:

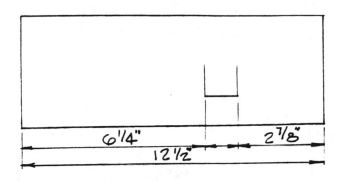

2. Determine the missing dimension in the member shown.

    Solution:

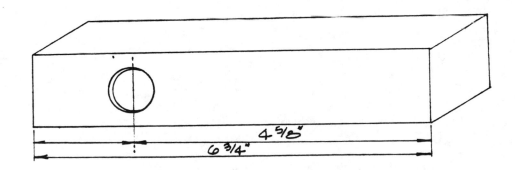

Find the missing dimensions in the part shown.

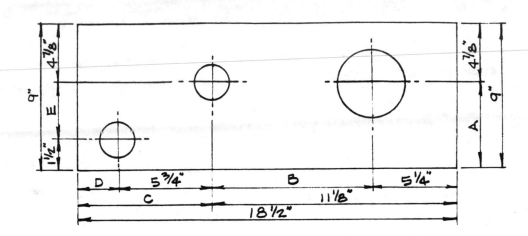

A ____

B ____

C ____

D ____

E ____

1. There are 117' of #5 rebar in stock. If 65' is required to weld a grating, how much #5 rebar remains in stock?

   Solution:

2. A welded steel tank requires 719 pounds of 1/8" steel plate. There are 1216 pounds of this material in stock. How much 1/8" steel plate is left in inventory after constructing the tank?

   Solution:

1. A fabrication order calls for 12 pieces of angle iron, each 7-1/2" long and 15 pieces of square tubing, each 9-1/4" long. What is the total length of angle iron required? What is the total length of square tubing required?

   Solution:

   Angle iron _____   Square Tubing _____

2. In the above problem, change the (12) pieces of angle iron to 11-1/4" long and the (15) pieces of square tubing to 6-5/8" long and proceed.

   Solution:

   Angle iron _____   Square tubing _____

1. Each piece of a tank support weighs 17 pounds. How much does each support weigh, using 7 pieces? What would be the total weight for an order of 30 supports?

   Solution:

2. A tank support requires 12 pieces of standard I-beam 92" long. If 7 supports are required, how many total inches would this be?

   Solution:

1. Nine welded legs each 18" long are cut from a stock length
    of material 16' long. Allow 3/16" for each cut. How many
    feet of material are used? How much material is left?

    Solution:

    Material used _____   Material left _____

2. In the above problem, change the 9 legs to 17-1/4" long
    and allow an additional 3/16" for each cut and proceed.

    Solution:

    Material used _____   Material left _____

1. Twelve pieces of square bar stock, each 18-1/2" long are cut from a piece of material 20 feet long. Allow 3/16" waste for each cut. How many feet of material are used? How much material is left?

   Solution:

   Material used _____   Material left _____

2. In the above problem, change the length of each cut piece to 17-3/4" and allow an additional 3/16" waste for each cut and proceed.

   Solution:

   Material used _____   Material left _____

1. How many pieces of angle iron 7-1/2" long can be cut from a stock piece 12' long? Allow 1/4" for each cut. How much is left?

   Solution:

   Material used _____   Material left _____

2. In the above problem, change each piece of angle iron to 8-1/8" long and proceed. (Allow an additional 1/4" for each cut.)

   Solution:

   Material used _____   Material left_____

1. Twenty-seven chemical tanks hold a total of 405 gallons.
    How many gallons does <u>each</u> tank hold?

    Solution:

2. If angle iron is purchased in 21' lengths. How many
    7' pieces can be sheared from 50 pieces of angle iron in
    stock?

    Solution:

1. How many 6-1/2" long pieces can be cut from a square tube 10'0" long?

    Solution:

    Pieces _____

2. How many 7-1/2" long pieces of angle can be cut from a stock piece 6'3" long?

    Solution:

    Pieces _____

1. How many 6'3" long pieces of #5 rebar can be cut from a
    stock piece 40'0" in length? How much rebar is left
    from the original piece?

    Solution:

    No. of pieces _____   Amount left _____

2. Five pieces of tubing 9-5/8" long are cut from a piece of
    tubing 8'0". Allow 1/8" for each cut. What is the total
    length removed? How much is left?

    Solution:

45

A piece of angle iron is divided into the fractional parts
shown. Write the decimal forms for the fractions indicated.
Also, determine the fractional dimension for G.

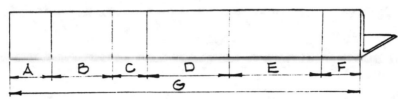

Solution:

A = <u>1/2"</u> = _____

B = <u>1-1/8"</u> = _____

C = <u>5/8"</u> = _____

D = <u>2-1/16"</u> = _____

E = <u>7/ 16"</u> = _____

F = <u>3/8"</u> = _____

G = _____

Five pieces of #4 rebar are found to be the following lengths. Change these measurements to improper fractions.

a. 7-1/2"   _____

b. 5-3/8"   _____

c. 9-1/4"   _____

d. 15-1/8"   _____

e. 5-3/16"

A piece of angle iron is cut into 8 parts. Name the fractional parts as indicated.

f. 1 part   _____

g. 3 parts   _____

h. 5 parts   _____

i. 7 parts   _____

Determine the missing dimensions in decimal form.

Solution:

A = 1.625"

B = 1.375"

C = _F - D_____

D = (2E)_____

E = 1.250"

F = 8'0"

1. How much square tubing is required to furnish 20 pieces 6.375" long?

    Solution:

2. A piece of pipe weighs 2.750 pounds. How many pieces can be cut from pipe weighing 82.5 pounds?

    Solution:

UNIT 7

BASIC ALGEBRA

In this unit, algebra will be used in the solution of problems that the welder will encounter both in training and on the job.

A typical problem is presented with step-by-step instructions to be used in working the solution. Similar problems to be worked using the step-by-step instructions are presented on pages that follow.

Proceed with caution and avoid short-cuts. Your confidence will build as you proceed. Remember speed is not an issue here; it will increase as your skills develop. As stated in the foreword, accuracy is foremost in importance.

Integers - an integer is a whole number eigher positive (+) or negative (-) in value as shown below on the number line:

$$\xrightarrow{\quad -8\ -7\ -6\ -5\ -4\ -3\ -2\ -1\quad 0\quad 1\ 2\ 3\ 4\ 5\ 6\ 7\ 8\quad}$$

The numbers to the right of '0' are positive and the numbers to the left are negative. Use the number line to tell which of two numbers is smaller. We know that 2 (or +2) is smaller than 4 (or +4). On the number line, 2 is to the left of 4. In general, the number to the left on the number line is the smaller.

Since 2 is to the left of 4, $(2 < 4)$. To write 2 is smaller than 4, use the symbol $<$. The point of the symbol "points" to the smaller number.

Also since -8 on the value line is to the left of 3 (or +3) it is smaller and may be designated $-8 < 3$, you can also write $3 > -8$ (3 is greater than -8).

Addition of integers - to add +2 and +4 on the number line, begin at 0 and draw an arrow 2 units to the right to represent +2; extend the arrow 4 units to the right to represent +4. The end of the arrow stops at +6, so that $+2 + (+4) = +6$.

To add +5 and -2 or $+5 + (-2)$ on the number line, drawn an arrow 5 units to the right of 0 to represent +5 and then draw an arrow 2 units to the left to represent -2. The answer from the number line is $+5 + (-2) = +3$.

Subtrations of integers - to subtract 3 from -2 written -2 -3, start at 0 on the number line and move 2 units to the left to represent -2, extend 3 units farther to the <u>left</u> representing subtracting -3, the result is -5, so that -2 -3 = -5.

The same answer would have resulted had we <u>added</u> -2 and -3, we would get -2 + (-3) = -5. "Subtracting" b from a always gives the same results as <u>adding</u> the <u>opposite</u> of b.

$$a - b + a + (-b)$$

Some examples:

$$-9 - 3 = -9 + (-3) = -12$$

$$-8 - 3 = -8 + (-3) = -11$$

$$4 - (-5) = 4 + (+5) = +9$$

$$-15 - 19 = -15 + (-19) = -34$$

$$11 - (-2) = 11 + (+2) = +13$$

When this occurs as 11 - (-2) change <u>both</u> negatives to <u>positive</u> and rewrite to 11 + (+2) = +13 (or just 13). These are known as "additive inverses" of each other or opposites.

<u>Multiplication of integers</u> - in general the products of two positive or two negative numbers is positive. The product of one negative number and one positive number is always negative.

Some examples:

$$(-9) \cdot (-3) = +27$$

$$(-9) \cdot (+3) = -27$$

$$(+9) \cdot (+3) = +27$$

$$(+9) \cdot (-3) = -27$$

<u>Division of integers</u> - the same rules apply as in multiplication.

Some examples:

$$\frac{-9}{-3} = +3 \qquad \frac{-27}{+3} = -9 \qquad \frac{-27}{-3} = +9$$

$$\frac{+9}{-3} = -3 \qquad \frac{+27}{+3} = +9 \qquad \frac{+27}{-3} = -9$$

<u>Order of operations</u> - when a problem involves several operations use the following order of operations:

50

1. Do any work inside parentheses
2. Work out any exponents
3. Start at the left and do any multiplications or divisions as you come to them.
4. Start again at the left and do additions or subtractions as you come to them.

Example:

$$(-5-1)\cdot(3)\div9+(4-2)\cdot(-5) = (-6)\cdot(3)\div9+2\cdot(-5)$$
$$= -18\div9+2\cdot(-5)$$
$$= -2+2\cdot(-5)$$
$$= -2+(-10)$$
$$= -12$$

or

$$9\cdot(6-10) = 9\cdot(-4)$$
$$= -36$$

Exponents are sometimes thought of as a "power" designation, as 7^2, the exponent 2 tells you to "square" the number by multiplying 7 by itself, 7 x 7 = 49.

Using an exponent shortens the problem as 5^3 means 5 x 5 x 5 = 125.

Roots - the square root of a number a, written $\sqrt{a}$, is the number that can be multiplied by itself to give the original number. As the $\sqrt{9}$ = 3 since 3 · 3 = 9. You may use this rule to find the square root of a number:

$$LOG \sqrt{a} = \frac{1}{2} \cdot LOGa$$

The cube root of a number a is written $\sqrt[3]{a}$. As $\sqrt[3]{8}$ = 2, since 2 x 2 x 2 = 8. You may use this rule to find the cube root of a number:

$$LOG \sqrt[3]{a} = \frac{1}{3} \cdot LOGa$$

or use your calculator.

An equation is an algebraic statement containing an unknown variable whose value must be found.

A linear equation has no variables with exponents.

As an example, x + 10 = 15 is a linear equation. To solve this equation use the addition rule. Adding the same number to both sides of the equation does not change it.

This equation x + 10 = 15 could be solved if x were alone on one side of the equals sign. To change x + 10 into just x, add -10 to both sides of the equation.

$$x + 10 = 15$$
$$x + 10 + (-10) = 15 + (-10)$$
$$\quad -10 \qquad\qquad -10$$
$$x \qquad\qquad = 5$$

Check your answer by substituting 5 for x in the original equation.

$$x + 10 = 15$$
$$5 + 10 = 15$$
$$15 = 15$$

Try this equation that contains x on both sides of the equals sign.

$$4x + 6 = 5x$$

To get x alone on one side add -4 to both sides.

$$4x + 6 + (-4) = 5x + (-4)$$
$$-4 \qquad\qquad -4$$
$$6 = 1x = x$$

Another equation (with fractions)

$$x + \frac{3}{8} = \frac{1}{16}$$

Add $\frac{-3}{8}$ to both sides

$$x + \frac{3}{8} + (\frac{-3}{8}) = \frac{1}{16} + (\frac{-6}{16})$$
$$\quad - \frac{3}{8}$$
$$x \qquad\qquad = \frac{1}{16} + (\frac{-6}{16}) \qquad (\frac{3}{8} = \frac{6}{16})$$
$$\qquad\qquad\qquad - \frac{6}{16}$$
$$x \qquad\qquad = - \frac{5}{16}$$

And another

$$25y - 6 = 24y + 9$$

$$25y + (-24y) - 6 = 24y + (-24y) + 9$$
$$-24y \qquad\qquad -24y$$
$$y \qquad -6 \qquad = \qquad +9$$
$$y \quad -6 + (+6) \quad = +9 +(+6)$$
$$y \qquad\qquad = 15$$

52

Discover that by isolating the variable on one side of the equation, it can be solved!

Not all equations can be solved by adding. To solve the equation $5x = 40$, we use the _multiplication_ rule.

Both sides of the equation may be multiplied by the _reciprocal_ of the value of the variables. The reciprocal of 5 is 1/5

$$(5 \text{ or } \frac{5}{1} = \frac{1}{5})$$

$$\text{as } 5x = 40$$

$$\frac{1}{5} \cdot 5x = \frac{1}{5} \cdot 40$$

$$x = \frac{40}{5}$$

$$x = 8$$

and another (ratio and proportion)

$5:100::x:500$ (this reads 5 is to 100 as x is to 500)

$$\frac{5}{100} = \frac{x}{500} \qquad \text{set up as fractions}$$

$$100x = 2500 \qquad \text{cross multiply}$$

$$\frac{1}{100} \cdot 100x = \frac{1}{100} \cdot 2500 \qquad \begin{array}{l} \text{multiply by reciprocal of 100} \\ \text{to get x alone} \end{array}$$

$$x = \frac{2500}{100}$$

$$x = 25$$

and another (using both addition and multiplication rule)

$$2x - 5 = 19$$

$$2x - 5 + 5 = 19 + 5 \quad \text{add 5 to both sides}$$
$$2x \qquad = 24 \qquad \text{multiply each side by reciprocal of 2}$$

$$\frac{1}{2} \cdot 2x \quad = \frac{1}{2} \cdot 24$$

$$x = \frac{24}{2}$$

$$x = 12$$

and another

$$8x - 6 = 4x + 18$$
$$8x - 6 + 6 = 4x + 18 + 6 \qquad \text{add 6 to both sides}$$
$$8x \qquad = 4x + 24$$
$$8x + (-4x) = 4x + (-4x) + 24$$
$$4x \qquad = 24$$

$$x = \frac{24}{4}$$

$$x = 6 \quad \frac{1}{4} \cdot 4x = \frac{1}{4} \cdot 24 \quad \text{multiply by reciprocal of 4}$$

Solve the following:

a. $3x + 6 = 24$

b. $5y - 9 = 36$

c. $6x = 54$

d. $x + 15 = 25$

e. $x + 1/2 = 5/8$

f. $10x - 5 = 9x + 8$

g. $y + 3/16 = 1/8$

1. A welder is preparing a sketch calling for a scale of
    1/4" = 1'0". If the dimension is 12'0", what length
    would it be drawn using the scale?

    Solution:

    Use ratio and proportion

    1. .25:1::x:12   (1/4" = .25)

    2. $\frac{.25}{1} = \frac{x}{12}$ set the proportion

    3. 1x = 3.00   cross multiply

    4. $\frac{1x}{1} = \frac{3.00}{1}$ divide both sides by 1

    5. x = 3"

2. In the above problem, change the dimension to 15'0" and
    proceed.

    Solution:

1. A welder is preparing a sketch calling for a scale of 1/2" = 1'0". If the dimension is 15'0", what length would it be drawn using the scale?

    Solution: (Remember 1/2" = .50)

2. In the above problem, change the scale to 1/8" = 1'0" and proceed.

    Solution: (Remember 1/8" = .125)

1. A drafter for Capitol Hill Drafting Services was preparing a drawing based on the following dimensions: 7'6", 9'3", 2'6" and 1'9". What do these dimensions total? If drawn at 1/8" = 1'0" scale, how long would the line be?

Solution:

1. Add the dimensions:

$$
\begin{array}{r}
7'6" \\
9'3" \\
2'6" \\
1'9" \\
\hline
19'24" = 21'0"
\end{array}
$$

Use ratio and proportion

2. .125:1::1:x:21 (.125 = 1/8")

$$\frac{.125}{1} = \frac{x}{21}$$ cross multiply
1x = 2.625

$$\frac{1x}{1} = \frac{2.625}{1}$$ divide each side by 1
x = 2.625 = 2-5/8"
(.625 = 5/8")

$$
\begin{array}{r}
2.625 \\
8\overline{)21.000} \\
16 \\
\hline
5\ 0 \\
4\ 8 \\
\hline
20 \\
16 \\
\hline
40
\end{array}
$$

2 = 2"   .625 = $\frac{5"}{8}$

2. In the above problem, change the scale to 1/4" = 1'0" and proceed to find the length of the line.

Solution:

58

1. In the previous problem, if the dimensions were changed to 9'6", 10'9", 3'6" and 1'3", what would be the total? If drawn at 1/4" = 1'0" scale, how long would the line be?

   Solution:

2. In the above problem, change the scale to 3/8" = 1'0" and proceed.

   Solution:

1. A line is 20" long and is drawn at 1/4" = 1'0" scale.
    How long would the line be at full scale?

    Solution:

    Use ratio and proportion

    a. .25:1::20:x   (say 1/4" is to 1' as 20" is to x)

    b. $\frac{.25}{1} = \frac{20}{x}$ set up the proportion

    c. .25x = 20   cross multiply

    d. $\frac{.25x}{.25} = \frac{20}{.25}$ divide both sides by .25

    $$.25\overline{)2000.}\ \ \overset{80.}{}$$
    $$\underline{200}$$
    $$0$$

    e. x = 80'

2. In the above problem, change the length of the line to
    16" and proceed.

    Solution:

1. A line is 12" long and is drawn at 1/4" = 1'0" scale.
    How long would the line be at full scale?

    Solution:

2. Change the above problem to a line 18" long and proceed.

    Solution:

1. A line has a dimension of 120.5' on a plot plan. When drawn to a scale of 1" = 20', how long would the line be?

   Solution:

   Use ratio and proportion

   a. 1:20::x:120.5   (1" is to 20' as x is to 120.5")

   b. $\dfrac{1}{20} = \dfrac{x}{120.5}$ set the proportion

   c. 20x = 120.5   cross multiply

   d. $\dfrac{20x}{20} = \dfrac{120.5}{20}$ divide both sides by 20

$$
\begin{array}{r}
6.025 \text{ or } 6\text{-}1/4" \\
20\,\overline{)120.5} \\
120 \\
\hline
50 \\
40 \\
\hline
100
\end{array}
$$

2. In the above problem, change the scale to 1" = 30' and proceed.

   Solution:

1. A line has a dimension of 225.5' on a plot plan. When drawn to a scale of 1" = 10', how long would the line be?

   Solution:

2. In the above problem, change the dimension to 275.0' and proceed.

   Solution:

A welder worked 42 hours @ the regular time of $12.00 per hour. What is this gross pay? (Rate of overtime is 1-1/2 times regular time.)

Solution:

1. 40 hours @ 12.00 = 40 x 12.00 = $480.00

$$\begin{array}{r} 12.00 \\ \underline{40} \\ 480.00 \end{array}$$

2. Time and one-half = 12.00 x $1\frac{1}{2}$ = 12.00 x $\frac{3}{2}$ = $\frac{36.00}{2}$ =

   $18.00 x 2 = $36.00 (2 hours overtime)

3. Gross pay = $480.00 + $36.00 = $516.00

   NOTE: Regular time = 40 hours

1. In the previous problem, if the regular rate of pay was $13.40 per hour, what would be his gross pay?

   Solution:

2. Change the rate of pay to $14.25 per hour and proceed.

   Solution:

UNIT 8

FOUR-SIDED FIGURES

Quadrilaterals are made up of squares and rectangles and other four-sided figures.

Squares

The formula for perimeter is P = 4 · s (s + s + s + s)

The formula for area is A = s · s (s^2)

Find the perimeter and area for the following square.

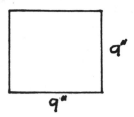

Solution:

1. P = 4 · s

2. P = 4 · 9

3. P = 36"

a. A = s · s

b. A = 9 · 9

c. A = 81 square inches

1. A welder was making a pattern that was square in shape and measured 12" on each side. Find the perimeter in linear inches and the area in square inches.

    Solution: Use the formulas $P = 4s$ and $A = s^2$

2. In the above problem, change the dimension of each side to 7" and proceed.

    Solution:

1. A welder was cutting some flat plate for a job. The plate was square in shape and measured 3' on each side. Find the perimeter in linear feet and the area in square feet.

   Solution:

2. In the above problem, change the dimension to 2'6" on each side and proceed. (First convert 2'6" to its decimal equivalent 2.5").

   Solution:

Rectangles

The formula for perimeter is P = 2l + 2w

The formula for area is A = lw.

Find the perimeter and area for the following rectangle:

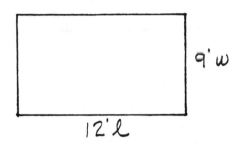

Solution:

1. P = 2l + 2w

2. P = 2(12) + 2(9)

3. P = 24 + 18

4. P = 42'

a. A = lw

b. 12 · 9

c. A = 108 sq. ft.

Study the formulas used above. Notice how they help you solve these problems. Without the use of these proven formulas, your task would be much more difficult.

Always follow through as shown substituting the values in the formulas.

1. In the preceding problem, change the dimensions to l = 30' and w = 12' and proceed.

   Solution:

2. In the preceding problem, change the dimensions to l = 19'6" and w = 7'9" and proceed.

   Solution:

1. A welding fabricator needs to cut some plate for a project. The size of the plate is to be 4' long and 2'6" wide. What is the perimeter and area of the plate? If the plate cost $.75 per square foot, what would be the cost for the plate? (Convert 2'6" to its decimal equivalent before proceeding.)

   Solution:

2. In the above problem, change the dimensions to 3'6" long and 1'9" wide and proceed.

   Solution:

Parallelogram

The formula for perimeter is $P = 2b + 2h$

The formula for area is $A = bh$

Find the perimeter and area of the following parallelogram:

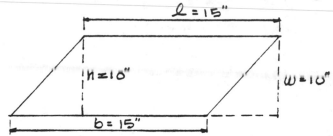

The rectangle $l = 15"$, $w = 10"$ has the same area and perimeter as the parallelogram shown.

Solution:

1. $P = 2b + 2h$

2. $P = 2(15) + 2(10)$

3. $P = 30 + 20$

4. $P = 50"$ (lineal inches)

a. $A = bh$

b. $A = 15 \cdot 10$

c. $A = 150$ in.^{2} (square inches)

1. A welder is preparing a pattern. It is shaped as a parallelogram of the following size:

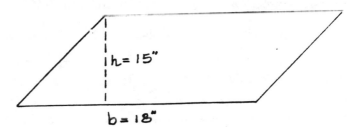

Find the perimeter.

Solution: P = 2b + 2h

2. In the above problem, change the dimension to b = 11" and
   h = 9-1/2" and proceed to find the perimeter and area.

Solution:

Trapezoid

The formula for area is A = 1/2 (b + B)h

Find the area of the following trapezoid:

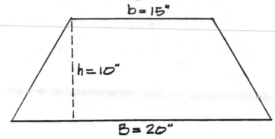

Solution:

1. A = 1/2 (b + B)h

2. A = 1/2 (15 + 20)10

3. A = 1/2 (35)10

4. A = 1/2 (350)

5. $A = \dfrac{350}{2}$

6. A = 175 in.^{2} (square inches)

1. Find the area in·square feet of the following trapezoid:

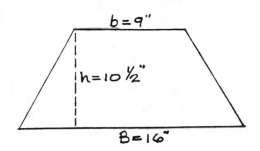

    Solution:

2. In the above problem, change the dimensions to B = 14",
    b - 11-1/2" and h = 8-1/2" and proceed to find the area.

    Solution:

CIRCLES

A circle is defined as a "figure bounded by a single curved line, every point of which is equally distant from the point at the center of the figure", or a flat circular figure where the exterior line is of equal distance from the center point.

The circumference of a circle is the measurement of the "distance around" at the outer edge. This measurement corresponds to the perimeter of a figure as square, rectangle or triangle.

The formula for finding the circumference of a circle is $c = \pi d$. The d represents the diameter of the circle, a straight line passing through the center point from one side to the other. The formula may also be written $c = 2\pi r$. The 'r' represents the radius which extends from the center in a straight line to the outer edge, therefore, 2 radii (2r) equal 1 diameter. Either use is correct. π has a mathematics "value" of 3.14 and is an irrational number.

In the formula $c = \pi d$, the first step is to substitute in the formula:

    c = π d (use d = 10")
    c = 3.14 x 10
    c = 31.4"

Circumference equals 31.4 inches

To find the area of a circle use this formula:

$A = \pi r^2$ (use r = 5")

Substitute in the formula.

$A = \pi r^2$

$A = 3.14 \times 5^2$

$A = 3.14 \times 25$

$A = 78.5 \text{ in.}^2$ (square inches)

To find the diameter of a circle with the known circumference, use this formula:

$d = \dfrac{c}{\pi}$ (use c = 31.4")

$$d = \frac{31.4}{3.14} \qquad d = 10"$$

To find the diameter of a circle with the known <u>area</u>, use this formula:

$$d = 2\sqrt{\frac{A}{\pi}} \qquad \text{(use A = 78.5 sq. in.)}$$

$$d = 2\sqrt{\frac{78.5}{3.14}}$$

$$d = 2\sqrt{25}$$

$$d = 10"$$

To find the length of an <u>arc</u> of a given circle, use this formula:

$$\text{Length of arc} = \frac{\text{Central angle x circumference}}{360°}$$

$$\text{L of arc} = \frac{C \text{ x angle x } C}{360} \qquad \text{(use a central angle of 30°}$$
$$\text{and a shaft diameter of 6")}$$

$$\text{L of arc} = \frac{30 \cdot 18.84}{360} \qquad (c = \pi d = 3.14 \cdot 6 = 18.84")$$

$$\text{L of arc} = \frac{565.2}{360}$$

$$\text{Length of arc} = 1.57"$$

To find the central angle if you know the length of the arc, use this formula:

$$\text{c. angle} = \frac{\text{L of arc x 360°}}{\text{circumference}} \qquad \text{(L of arc = 1.57"}$$
$$c = 18.84")$$

$$\text{c. angle} = \frac{1.57 \text{ x } 360}{18.84}$$

$$\text{c. angle} = \frac{565.2}{18.84}$$

$$\text{c. angle} = 30°$$

1. Find the circumference and area of a plate with a <u>radius</u> of 3 feet.

   Solution:

2. A carpenter is building a circular gazebo with a diameter of 20'. A welder is preparing an estimate for a W.I. railing. Find the circumference.

   Solution:

   Use this formula: C = πd to find circumference.

1. In the preceeding problem, change the diameter of the
    gazebo to 28' and proceed with the solution.

    Solution:

2. A welder needs to find the area in square inches of the
    end of a shaft 2-3/4" in diameter.

    Solution:

    1. Use this formula for area $A = \pi r^2$

    2. $A = \pi r^2$

    3. $A = 3.14 \cdot 1.375^2$

    4. $A = 5.9365$ square inches (area)

1. In the preceding problem, change the shaft diameter to 2-1/4" and proceed with the solution.

   Solution:

2. A machinist must cut the keyway shown in the figure below:

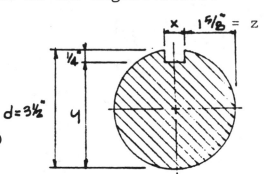

   Solution:

   1. Find the width of the keyway (x)

      x = 3-1/2"d - 2(1-5/8")

      x = 3-1/2" - 3-1/4"

      x = 1/4" (width of keyway)

   2. Find the dimension y

      y = 3-1/2"d - 1/4"

      y = 3-1/4"

1. In the preceding problem, the diameter of the shaft is
   3-1/4". Solve for x and y. The depth of keyway is 1/4" and
   offset dimension z = 1 1/2"

   Solution:

2. A welder must find the <u>diameter</u> of a shaft with a circumference of 14".

   Solution:

   Use this formula and substitute

   1. $d = \dfrac{c}{\pi}$

   2. $d = \dfrac{14}{3.14}$

   3. d = 4.458" or approx. 4-29/64"

1. In the preceding problem, change the circumference to 11" and find the diameter.

   Solution:

2. A plate base has an area of 28.26 square inches. Find the diameter.

   Solution:

   Use this formula and substitute

   1. $d = 2 \sqrt{\dfrac{A}{\pi}}$

   2. $d = 2 \sqrt{\dfrac{28.26}{3.14}}$

   3. $d = 2 \cdot \sqrt{9}$

   4. $d = 6"$ (diameter of plate base)

1. In the preceding problem, the area of the plate base
   is changed to 50.24 square inches. Find the diameter.

   Solution:

2. If the area of the plate base is 47.1 square inches, what
   would be the diameter?

A welding fabricator is preparing an estimate to fabricate a 6'0" high W.I. fence for the playing field shown below. Find the circumference.

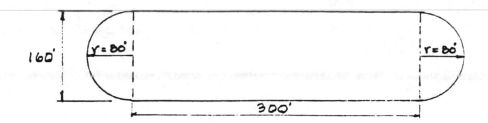

Solution:

1. Find the lengths of the <u>sides</u> of the rectangle.

2. 2 (1) = 2 · 300 = 600'

3. Find the circumference of 2 half circles (1 full circle) with the formula C = π d.

    C = π d
    C = 3.14 x 160
    C = 502.4'

4. Add 600' + 502.4' to get total feet of 1102.4' or 1102-4" (4/12 = 1/3 of a foot or 4")

1. In the preceding problem, find the circumference of the playing field if the length was 400' and the width was 210' with a radius of 105' at each end.

   Solution:

2. In the previous problem, change the width only to 190' and proceed.

   Solution:

CHORDAL SEGMENT OF A CIRCLE

Find the arc (L), chord (C), angle (A), radius (R) and height (H) for the segment of the circle shown.

L = Length of Arc
C = Length of Chord
A = Central Angle
R = Radius of Circle
D = Diameter (2R)
H = Height or Depth of
     Segment

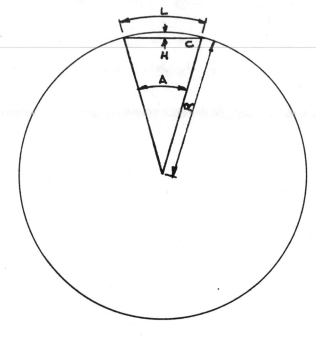

Find L

Given: Size of Angle (A) and Diameter (D)
Example - Use 30° angle and diameter of 3"

Solution:

1. Find circumference of circle

$$C = \pi D, \quad C = 3.14 \times 3 = 9.42"$$

2. Use formula $L = \dfrac{Cir. \times A}{360}$

3. $L = \dfrac{9.42 \times 30}{360}$

4. $L = 9.42 \times \dfrac{1}{12} = \dfrac{9.42}{12} = .785"$ (or approx. 25/32")

5. $L = .785"$

or

Use this formula $L = 0.01745 \times R \times A$
                     (constant)

86

a.   L = 0.01745 x 1.5 x 30

b.   L = 0.01745 x 1.5 = 0.026175

c.   L = 0.026175 x 30 = .785"

d.   L = .785"

Find R

Given: Length of arc (L) and size of angle (A)
Example - use L = .785" and A = 30°

Solution:

1.   Use this formula $R = \dfrac{\overset{(constant)}{57.296 \times L}}{A}$

2. $R = \dfrac{57.296 \times .785}{30}$

3. $R = \dfrac{44.97736}{30}$

4.   R = 1.499 or round to 1.5"

or

Use this formula $d = \dfrac{c}{\pi}$

a. $d = \dfrac{9.42}{3.14}$

b.   d (dia.) = 3"

c. $R = \dfrac{dia.}{2}$ or $\dfrac{3}{2} = 1\frac{1}{2}$ or 1.5"

Find A

Given: length of arc (L) and radius (R)
Example - use L = .785" and R = 1.5"

Solution:

1.   Use this formula $A = \dfrac{\overset{(constant)}{57.296 \times L}}{R}$

2. $A = \dfrac{57.296 \times .785}{1.5}$

3. $A = \dfrac{44.97736}{1.5}$

4. $A = 29.9849$ round to 30

5. $A = 30''$

or

Use this formula $A = \dfrac{L \times 360°}{C}$

a. $A = \dfrac{.785 \times 360}{9.42}$

b. $A = \dfrac{282.6}{9.42} = 30$

c. $A = 30°$

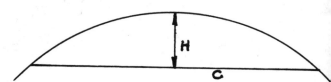

Find H

Given: Radius (R) and chord (C)
Example - R = 1.5" and c = .7764 (constant for 30° =
.2588 x dia. 3" =
.7764)

Solution:

1. Use this formula $H = R - 0.5\sqrt{4R^2 - C^2}$

2. $H = 1.5 - 0.5\sqrt{4 \cdot 1.5^2 - .7764^2}$

3. $H = 1.5 - 0.5\sqrt{4 \cdot 2.25 - .60279}$

4. $H = 1.5 - 0.5\sqrt{9 - .60279}$

5. $H = 1.5 - 0.5\sqrt{8.39721}$

6. $H = 1.5 - 0.5 \times 2.8977$

7. $H = 1.5 - 1.448$

8. $H = 0.052''$ or approx. 3/64"

Find C (another way to find chord C without using a constant)

Given: Radius (R) and Height (H)
Example - R = 1.5" and H = 0.052"

Solution:

1. Use the formula $C = 2\sqrt{H(2R-H)}$

2. $C = 2\sqrt{0.052(2 \cdot 1.5 - 0.052)}$

3. $C = 2\sqrt{0.052(3 - 0.052)}$

4. $C = 2\sqrt{0.052 \cdot 2.948}$

5. $C = 2\sqrt{.153296}$

6. $C = 2 \times .39153$

7. C = 0.783 or approx. 25/32"

   NOTE: The chord (using constant for 30°) was .7764 , using the above formula C equaled .783 or a difference of .0066 which is insignificant for low tolerance work.

TABLE OF CHORD CONSTANTS

DEGREE	CONSTANT
120°	0.8660
90°	0.7071
60°	0.5000
45°	0.3827
30°	0.2588
20°	0.1736
15°	0.1305

Constants for degrees not shown can be determined by multiplying the constant per degree of 0.00825 (approx.) by the degree of angle.

Based on 1" Diameter

1. Find length (L) of arc in a circle segment with a central angle of 20° and a·diameter (D) of 3-1/2".

   Solution: Use either method

2. Find length (L) of arc in a circle segment with a central angle of 40° and a diameter (D) of 8".

   Solution: Use either method

1. Find radius (R) with the length of arc (L) of .950"
   and central angle (A) of 25°.

   Solution: Use either method

2. Find radius (R) with the length of arc (L) of .8" and
   central angle of 20°.

   Solution: Use either method

1. Find angle (A) with the length of arc (L) of 1.125"
    and radius (R) or 1.750".

    Solution: Use either method

2. Find angle (A) with the length of arc (L) of 1.375" and
    radius (R) of 1.5".

    Solution: Use either method

1.   Find height (H) for the chord segment for Problem #1
      on page 90 .

      Solution:

2.   Find height (H) for the chord segment for Problem #2
      on page 91 .

      Solution:

1. Find the length of chord (C) for Problem #1 on page 90.

    Solution:

2. Find the length of chord (C) for Problem #2 on page 90.

    Solution:

UNIT 10

VOLUME

Volume is best defined as the cubic content of a space occupied in three dimensions, as length, width and height (or depth). Volume is (then) three dimensional.

The formula used to find the volume of a box or cube is V = lwh (l = length, w = width, h = height).

Example--find the volume of the box shown here.

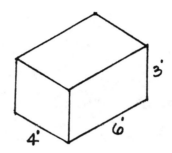

Solution:

V = LWH

V = 6 · 4 · 3

V = 72 ft.^{3} (cubic feet)

1. Find the volume in cubic feet of a metal tank 12' long, 6' wide and 4' high.

    Solution:

    Use the formula $V = lwh$ to find the volume and substitute values.

2. Find the volume in cubic feet of a metal tank 16' long, 7'6" wide and 3'9" high.

    Solution:

A tank manufacturer is fabricating a sludge tank to hold 500 gallons. How many <u>cubic</u> <u>feet</u> must the tank hold?

Solution:

1. There are 7.48 gallons per cubic feet.

2. Cubic feet = 500 ÷ 7.48

3. c.f. = 66.84 or round to 67 c.f.

The tank is to be 5 feet deep.

a. To determine cubic feet required for a 5' deep tank, use this formula:

b. c.f. = $\frac{67}{5}$ = 13.4

c. $\sqrt{13.4}$ = 3.66 x 3.66 or 3'8" x 3'8"

d. Therefore, a tank 3'8" x 3'8" x 5' deep equals 67 cubic feet multiplied by 7.48 equals 500 gallons.

1. In the preceding problem, change the requirements to a
    600 gallon tank and proceed with the solution.

    Solution:

2. Change the requirements to a 425 gallon tank and proceed
    with the solution.

    Solution:

A cylindrical tank is to be fabricated to hold 500 gallons
of sludge. There are 7.48 gallons per cubic foot, therefore,
the tank must hold approx. 67 cubic feet. If the tank is to
be 5' in diameter, how long must it be?

(500 ÷ 7.48 = 66.84 round to 67 c.f.)

Solution:

1. The formula is $V = \pi r^2 h$ for volume

2. $\pi r^2 = 3.14 \cdot 2.5^2$ or $3.14 \cdot 6.25 = 19.625$ sq. ft.

3. Therefore, 67 c.f. ÷ 19.625 = 3.41' or round to 3.5

4. length of $h = \dfrac{67}{3.14 \cdot 2.5^2}$

5. $h = \dfrac{67}{19.625}$

6. h = 3.41' round to 3.5 or 3'6"

7. Therefore, a cylindrical tank 5' in diameter must be
    approx. 3'6" long to carry 500 gallons.

1. In the preceding problem, change the tank diameter to
    4' and proceed with the solution.

    Solution:

2. Change the tank diameter to 7'6" and proceed with the
    solution.

    Solution:

1. A tank with a diameter of 4' and 7' long holds how many cubic feet? If 3% of the capacity is lost to evaporation, how many cubic feet are left?

Solution:

1. Use this formula for volume

   $V = \pi r^2 h$ and substitute

2. $V = \pi r^2 h$

3. $V = 3.14 \cdot 2^2 \cdot 7$

4. $V = 3.14 \cdot 4 \cdot 7$

5. $V = 87.92$ cubic feet

To find the number of cubic feet left after evaporation, use ratio and proportion.

a. $87.92:100::x:97$

b. $\dfrac{87.92}{100} = \dfrac{x}{97}$ (number of cubic feet left)
   (97% left after evaporation)

c. $100x = 8528.24$ cross multiply

d. $\dfrac{100x}{100} = \dfrac{8528.24}{100}$ divide both sides by 100

e. $x = 85.28$ cubic feet left after 3% is lost to evaporation.

1. <u>Cylinders</u>--the formula for finding the volume of a right
   <u>cylinder</u> is:

   $V = \pi r^2 h$

   Find the volume of a can of condensed milk with a radius
   of 1.5" (r) and 4.5" tall (h).

   $V = \pi r^2 h$ (πr^2 = area, then multiply by height)

   $V = 3.14 \times 1.5^2 \times 4.5$

   $V = 3.14 \times 2.25 \times 4.5$

   $V = 7.065 \times 4.5$

   $V = 31.79 \text{ in.}^3$ (cubic inches)

2. In the above problem, find the volume of a can of pop with
   a <u>diameter</u> of 2-1/2" and 4-3/4" tall (h).

   Solution:

1. In the preceding problem, find the volume in <u>pints</u> of the can of condensed milk.

Solution:

There are 8 pints in one gallon.

One gallon equals 231 cubic inches

One pint equals $231 \div 8$ or 28.9 in.^{3} (cubic inches)

1. Use this formula

2. $V = \dfrac{\pi r^2 h}{28.9}$

3. $V = \dfrac{3.14 \cdot 1.5^2 \cdot 4.5}{28.9}$

4. $V = \dfrac{3.14 \cdot 2.25 \cdot 4.5}{28.9}$

5. $V = \dfrac{7.065 \cdot 4.5}{28.9}$

6. $V = \dfrac{31.79}{28.9}$

7. $V = 1.1$ pints or 1-1/10 pints

2. In the above problem, determine how many liters (1) in the can of condensed milk.

Solution:

103

1. A tank manufacturer builds tanks for petroleum companies. One such tank is 24' long with a radius of 6'. Find the volume in cubic feet.

Solution:

1. Use this formula and substitute

2. $V = \pi r^2 h$

3. $V = 3.14 \cdot 6^2 \cdot 24$

4. $V = 3.14 \cdot 36 \cdot 24$

5. $V = 113.04 \cdot 24$

6. $V = 2712.96$ ft.^{3} or round off to 2713 cubic feet

2. In the above problem, find the volume in cubic feet and in gallons if the tank had a diameter of 10' and was 20' in length. (One cubic foot = 7.48 gallons)

To find the volume of a <u>Right Circular Cone</u>, use this formula:

$V = \frac{1}{3} Ah$ and substitute.

Solution:

Use a base <u>diameter</u> of 6" and altitude (h) of 12"

1. $V = \frac{1}{3} Ah$

2. A in the formula represents area and should be determined first

3. $A = \pi r^2$

4. $A = 3.14 \cdot 3^2$

5. $A = 3.14 \cdot 9 = 28.26$ sq. in. (in^2)

a. $V = \frac{1}{3} Ah$

b. $V = \frac{1}{3} (28.26)12$

c. $V = \frac{339.12}{3}$

d. $V = 113.04$ in.^{3} (cubic inches)

1. In the preceding problem, change the base diameter to 8" and altitude to 16" and proceed with the solution.

   Solution:

2. In the preceding problem, change the base diameter to 10" and altitude to 20" and proceed.

   Solution:

Spheres--the formula for finding the volume of a sphere is:

$$V = \frac{4}{3}\pi r^3$$

Find the volume of a toy balloon with a _radius_ of 6".

Solution:

1. $V = \frac{4}{3}\pi r^3$

2. $V = \frac{4}{3}(3.14)6^3$

3. $V = \frac{4}{3}(3.14)216$

4. $V = \frac{4}{3} \cdot 678.24$

5. $V = \frac{2712.96}{3}$

6. $V = 904.32$ (round off to 904 cubic inches)

1. In the preceding problem, find the volume of a tank float with a <u>diameter</u> of 6".

    Solution:

2. In the above problem, change the float <u>diameter</u> to 4-1/2" and proceed.

    Solution:

UNIT 11

SIMILAR FIGURES

Figures that are alike in shape are known as "similar figures". Squares and circles fall into this category. A square is similar to any other square, etc. A ratio exists between corresponding parts of similar figures. However, not all triangles are similar since they may have different shapes.

The two squares shown here are similar figures.

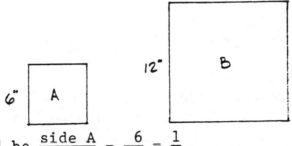

The ratio would be $\dfrac{\text{side A}}{\text{side B}} = \dfrac{6}{12} = \dfrac{1}{2}$

The length of the sides have a ratio 1/2, their <u>areas</u> will have the ratio $1/2^2$ or 1/4 or area A equals = $(\text{side A})^2$ and area B = $(\text{side B})^2$.

Compare the <u>areas</u> of the above squares.

$\dfrac{\text{Side B}^2}{\text{Side A}^2} = \dfrac{12^2}{6^2} = \dfrac{144}{36} = \dfrac{4}{1}$

1. Two patterns used in the welding shop are similar figures as shown here.

    Solution:

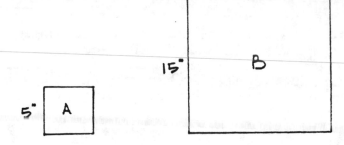

$$\frac{\text{Side A}}{\text{Side B}} = \frac{5''}{15''} = \frac{1}{3}$$

The length of the sides have the ratio 1:3 or $\frac{1}{3}$

Their <u>areas</u> have the ratio $\frac{12}{3}$ of $\frac{1}{9}$ or 1:9

$$\frac{\text{Side B}}{\text{Side A}} = \frac{15^2}{5^2} = \frac{225}{25} = \frac{9}{1}$$

1. In the preceding problem, change side A dimension to 6"
   and side B dimension to 18". Solve for length of sides
   ratio and areas ratio.

   Solution:

2. In the above problem, change side A dimension to 4" and
   side B dimension to 12". Solve for length of sides ratio
   and areas ratio.

   Solution:

1. The two circles shown here are similar figures.

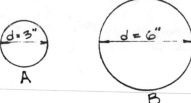

A

B

The <u>areas</u> of these circles are directly proportional to
the <u>squares</u> of their <u>diameters</u>.

$$\frac{B}{A} = \frac{d^2}{d^2} = \frac{6^2}{3^2} = \frac{36}{9} = \frac{4}{1}$$

NOTE: In the above examples (squares and circles) when
the side or diameter is doubles (1:2 ratio),
the <u>area</u> increases 4 times (4:1 ratio)

2. In the above problem, change A to d = 9 and B to d = 18
and proceed.

Solution:

1. A steel pulley with a diameter of 7" is similar to a pulley with a diameter of 21". What is the ratio of the diameters and the areas?

   Solution:

   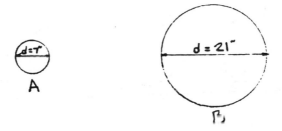

   $\dfrac{\text{Circle A}}{\text{Circle B}} = \dfrac{7"d}{21"d} = \dfrac{1}{3}$ or 1:3 ratio of diameters

   Their areas have the ratio $\dfrac{1}{3}^2$ or $\dfrac{1}{9}$ or 1:9 (9:1)

   $\dfrac{\text{Area Circle B}}{\text{Area Circle A}} = \dfrac{21^2}{7^2} = \dfrac{441}{49} = \dfrac{9}{1}$ or 9:1 ratio of areas

   The area of Circle B is 9 times that of Circle A.

2. In the above problem, change A to 8" diameter and B to 24" diameter and proceed.

   Solution:

1. The two triangles shown here are similar in shape with
   corresponding angles. Though corresponding angles are
   equal, corresponding sides need not have the same length,
   however the ratio of the length of corresponding sides
   is always the same.

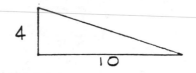

Solution:

Use ratio and proportion

The <u>ratio</u> would be

1. 4:2::10:x

2. $\frac{4}{2} = \frac{10}{x}$ cross multiply

3. 4x = 20

4. $\frac{4x}{4} = \frac{20}{4}$ divide both sides by 4

5. x = 5

6. Therefore, the proportion of 4:2 = 10:5, or multiply
    both sides by the reciprocal of 4.

a. $\frac{1}{4} \cdot 4x = \frac{1}{4} \cdot 20$

b. $x = \frac{20}{4}$

c. x = 5

NOTE: As shown above similar triangles can be used to
       find heights of objects and distances between
       points.

1.   Solve for proportion in these similar triangles.

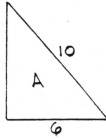

Solution:

The ratio would be 10:5::6:x

Solve for <u>proportion</u> (x)

2.   In the above problem, change A dimensions to 14 and 8
     and B dimension to 7 and solve for x.

     Solution:

1. In the preceding problem, change dimensions on triangle
    A to 8 and 12 and triangle B to 6 and x and proceed
    with solution.

    Solution:

2. In the above problem, change A dimensions to 10 and 20
    and B dimension to 10 and solve for x.

    Solution:

UNIT 12

ANGLES AND TRIANGLES

1. Angles are measured in degrees. There are 360° around a <u>vertex</u>. Degrees are written with the degree symbol ° (360 degrees is written 360°).

   A vertex is the point from which two lines extend to form an angle.

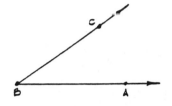

Angle ABC or CBA may be called by its vertex only. The angle can be called angle B.

Study the following angles and identify them accordingly.

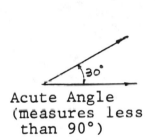

Acute Angle
(measures less
than 90°)

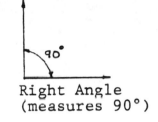

Right Angle
(measures 90°)

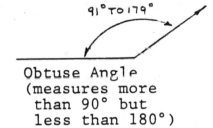

Obtuse Angle
(measures more
than 90° but
less than 180°)

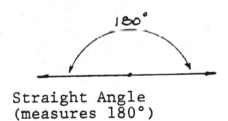

Straight Angle
(measures 180°)

117

2. Triangles:

The sum of the interior angles in any triangle is 180°.
Find the value of x.

Solution:

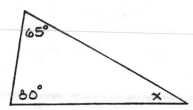

Let x represent the missing angle. Since the sum of
the angles in a triangle equal 180°, we have this equation:

1.   80 + 65 + x = 180

2.   145   + x = 180

3.   x = 180 - 145

4.   x = 35

                    or

a. To solve this equation, add -145 to both sides

b. 145 + x + (-145) = 180 + (-145)
    -145   -145

c.   x   = 35

d. By adding -145 to both sides we get x alone on one
    side. To prove, substitute 35 for x in the equation.

    145 + 35 = 180 (angle x = 35°)

    Both sides equal 180

Triangles:

The sum of the interior angles in any triangle is 180°. Find the value of x.

Solution:

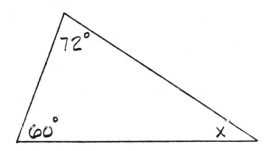

Let x represent the missing angle. Since the sum of the angles in a triangle equal 180°, we have this equation:

1. $72 + 60 + x = 180$

2. $132 + x = 180$

3. $x = 180 - 132$

4. $x = 48$

or

a. To solve this equation, add -132 to both sides.

b. $132 + x + (-132) = 180 + (-132)$
    -132 -132

c. $x = 48$

d. By adding -132 to both sides we get x alone on one side. To prove, substitute 48 for x in the equation.

$132 + 48 = 180$ (angle x = 48°)

Both sides equal 180

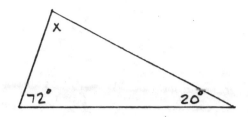

1. Let x represent the missing angle.

   Solution:

2. Let x represent the missing angle.

   Solution:

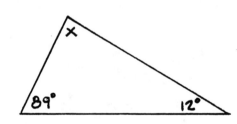

1. The exterior angle of the triangle:

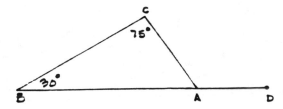

Solution:

1. The exterior angle CAD equals the sum of the two opposite interior angles (angles B and C).

2. The exterior angle CAD = 30 + 75 or 105°

2. Example: Find the value of x in the following triangle:

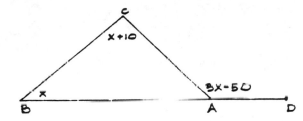

Solution:

The exterior angle CAD equals the sum of interior angles B and C. This gives the equation:

a. $\qquad$ x + (x + 10) = 3x - 50

b. simplify $\qquad$ 2x + 10 = 3x - 50

c. add +50 to both sides
$$2x + 10 + (+50) = 3x - 50 + (+50)$$

d. $\qquad$ 2x + 60 = 3x

e. add -2x to both sides
$$2x + 60 + (-2x) = 3x + (-2x)$$
$$-2x \qquad\qquad -2x$$
$$60 = x$$

f. substitute for x. Both sides of the equation = 130°

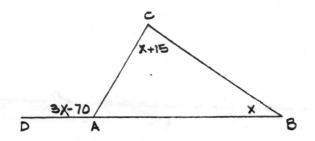

1. Find the value of x.

    Solution:

2. Find the value of x.

    Solution:

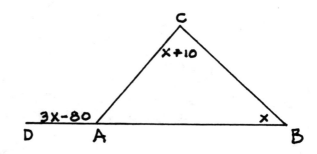

The right triangle:

The Pythagorean formula is $a^2 + b^2 = c^2$ and applies only to right triangles. Use your calculator or square root tables to find the following:

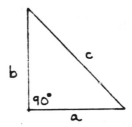

Example: Let a = 6, b = 8

Use the formula $a^2 + b^2 = c^2$ and substitute

$$6^2 = 8^2 = c^2$$
36 + 64 = 100, therefore, $100 = c^2$
the square root of 100 = 10
side c = 10

To find side b if $c^2 = 100$ and $a^2 = 36$

set up this equation: $b^2 = c^2 - a^2$

$$b^2 = 100 - 36$$

$$b^2 = 64 \quad b = 8$$

To find side a if $c^2 = 100$ and $b^2 = 64$

set up this equation: $a^2 = c^2 - b^2$

$$a^2 = 100 - 64$$

$$a^2 = 36 \quad a = 6$$

To find side b if $c^2 = 100$ and $a^2 = 36$

$$a^2 + b^2 = c^2$$

$$6^2 + b^2 = 10^2$$

$$36 = b^2 = 100$$

Add -36 to both sides

$$36 + b^2 + (-36) = 100 + (-36)$$

$$-36 \qquad\qquad -36$$

$$b^2 \qquad\qquad = 64 \qquad b = 8$$

To find side a if $b^2 = 64$ and $c^2 = 100$

$$a^2 + b^2 = c^2$$

$$a^2 + 8^2 = 10^2$$

$$a^2 + 64 = 100$$

Add -64 to both sides

$$a^2 + 64 + (-64) = 100 + (-64)$$

$$-64 \qquad\qquad -64$$

$$a^2 = 36 \qquad a = 6$$

To find side c if $a = 6$ and $b = 8$

Use the formula: $a^2 + b^2 = c^2$ and substitute

$$36 + 64 = c^2$$

$$100 = c^2 \qquad c = 10$$

Refer to page 123

Solution:

1. Find the value of c if a = 12 and b = 16

2. Find the value of b if c = 5 and a = 3

3. Find the value of a if c = 40 and b = 32

What is the length of side c of the right triangle shown?

    Solution:

1.   Use the formula: $a^2+b^2=c^2$
                                     (Pythogorean Formula)

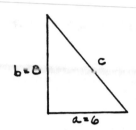

                                     then $\sqrt{c^2}$ = c

                                     $a^2 = 36$ (6x6)

                                     $b^2 = 64$ (8x8)

                                     $c^2 = 100$

2. Side c - 10"

                                   or

a. $c = a^2 + b^2$ $(a^2 = 36, b^2 = 64)$

b. c = 36 + 64

c. $c = \sqrt{100}$

d. c = 10

1. In the previous problem, if the lengths of side b and side a were change to 3" and 4", what would be the length of side c?

   Solution:

2. In the above problem change side b to 9" and side a to 7" and proceed.

   Solution:

A worker needs to find the dimension of the slope of the roof of this house from the ridge to the wall.

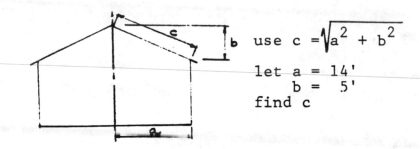

use $c = \sqrt{a^2 + b^2}$

let $a = 14'$
$ b = 5'$
find c

Solution:

1. $c = \sqrt{a^2 + b^2}$ or $c^2 = a^2 + b^2$

2. $c = \sqrt{196 + 25}$

3. $c = \sqrt{221}$ = 14.86' or c = approx. 14'-10-1/4"

1. A worker needs to find the dimension of the slope of the roof from the ridge to the wall.

   See preceeding illustration.

   let a = 6'
   let b = 18'
   find c

   Solution:

2. In the above problem, change a to 12' and b to 6' and proceed.

   Solution:

1. A welder needs to make a pattern which he will use to cut gussets from 1/4" steel plate. He needs to determine the length of side c.

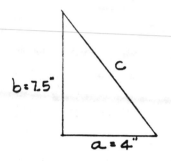

Solution:

1. $a^2 + b^2 = c^2$

2. $4^2 + 7.5^2 = c^2$

3. $16 + 56.25 = 72.25$, therefore, $72.25 = c^2$

4. $\sqrt{72.25} = 8.5$ side c = 8.5"

2. In the above problem, change a to 7" and b to 12" and proceed.

Solution:

1.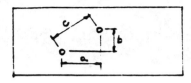

A welder needs to find the missing dimension.   Let a = 4, b = 3

Solution:

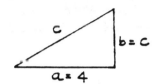

1. $a^2 + b^2 = c^2$

2. $4^2 + 3^2 = c^2$

3.   16 + 9 - 25, therefore $c^2 = 25$

4. $\sqrt{25}$ = 5, side c = 5"

2.   In the above problem, change b to 9" and a to 7" and proceed.

Solution:

1. A road builder needs to determine the length of the slope of the following area to be filled.

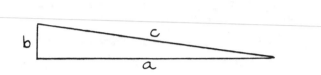

let 40' = a
    10' = b
find c

Solution:

Use your calculator to find the square root for c

1. $c = \sqrt{a^2 + b^2}$ or $a^2 + b^2 = c^2$

2. $c = \sqrt{1600 + 100}$

c. $c = \sqrt{1700}$ = 41.23' or c = approx. 41'-2-3/4"

2. In the above problem, change to a piece of fabricated plate that is 10' long and 6' high with a sloping side. Find the length of the slope.

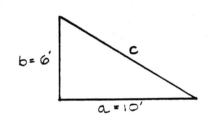

Solution: Use the formula $c \sqrt{a^2 + b^2}$

1. A road builder needs to determine the length of the slope of the following area to be filled.

let a = 50'
   b = 8'
find c

Solution:

2. In the above problem change a to 60' and b to 12' and proceed.

Solution:

UNIT 13

PERCENTAGES

This unit will present problems and solutions covering percentages. There are different ways to solve percentage
problems but the two most widely used are the "business"
method and the "algebraic" method. A third method, the
"direct ratio" method will also be presented.

All problems presented in this unit will explore the use of
these methods, so you can make the decision as to which
method best suits you.

A "whole" equals 100% as illustrated in the example shown
here. This graphic might represent a breakdown of operating
expenses for a small company. The total of the percentages
shown equal 100%.

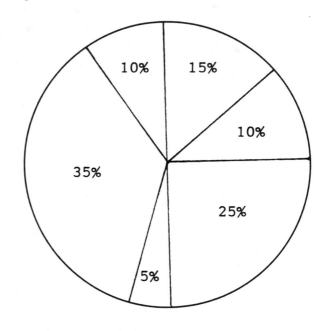

```
 35% = wages
 10% = office expense
 15% = overhead
 25% = marketing
 10% = gross profit
  5% = profit sharing
100%
```

It is important to remember that a given percentage represents
that part of the whole.

If the above company has gross sales of $1,575,000.00 for the
year then 35% of that amount would cover wages.

    35% of $1,575,000.00 = .35 x 1,575,000.00 or $551,250.00
    for wages.

A percentage is a given rate or proportion in every hundred.

1. Percentage formulas are outlined as follows:

P = Part of Base   <u>Business</u> <u>Method</u>
R = Rate as %
B = Base

To avoid confusion <u>always</u> set up this "pie"

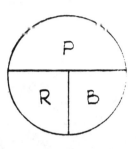

Examples: (let P = 5, R = 10%, B = 50)

a. P (part) = R (Rate) x B (base)
    P = RB
    P = .10 · 50 (10% x 50)

$$\begin{array}{r} 50 \\ .10 \\ \hline 5.00 \end{array}$$

    P = 5

The answer 5 is 10% of the base 50

b. R (rate) = $\frac{P \text{ (part)}}{B \text{ (base)}}$

    R = $\frac{5}{50}$

$$50 \overline{)\begin{array}{l} .1 = 10\% \\ 5.0 \\ \underline{5\ 0} \end{array}}$$

    R = .10 or 10%

The answer is the rate of 10%

c. $B \text{ (base)} = \dfrac{P \text{ (part)}}{R \text{ (rate)}}$

$B = \dfrac{5}{.10}$

$$.10. \overline{)5.00.} \quad \begin{array}{r} 50. \\ \underline{50} \\ 00 \end{array}$$

$B = 50$

The answer 50 is the base

2. Ratio and proportion may be used as follows:

To find the <u>rate</u>   <u>Algebraic Method</u>

a. x:100::5:50 is the ratio   5 = part
(this says x is to 100 as   50 = base
   5 is to 50)   x = rate

Therefore the proportion

(Rate) $\dfrac{x}{100} = \dfrac{5}{50}$ cross multiply
                        50x = 500 or 500 = 50x

*
See Page 138 $\dfrac{5\cancel{0}x}{\cancel{5}0} = \dfrac{500}{50} = 10$ divide both sides by 50
                           x = 10

Threrfore, $\dfrac{10}{100} = \dfrac{1}{10}$ or .10 or 10%

Substitute 10 for x in the equation

$(\dfrac{10}{100} = \dfrac{5}{50})$ cross multiply
             500 = 500
             both sides are equal

b. To find the part

10:100::x:50 is the ratio

Therefore the proportion

$\frac{10}{100} = \frac{x}{50}$ (part)   cross multiply
                                          100x = 500

$\frac{100x}{100} = \frac{500}{500} = 5$ divide both sides by 100
                                          x = 5

Substitute 5 for x in the equation

Therefore $\frac{10}{100} = \frac{5}{50}$ cross multiply
                                          500 = 500
                                          both sides are equal

c. To find the base

10:100::5:x is the ratio

Therefore the proportion

$\frac{10}{100} = \frac{5}{x}$ (base)   cross multiply
                                    10x = 500

$\frac{10x}{10} = \frac{500}{10} = 50$ divide both sides by 10
                                    x = 50

Substitute 50 for x in the equation

Therefore $\frac{10}{100} = \frac{5}{50}$ cross multiply
                                    500 = 500
                                    both sides are equal

* $50x = 500$

Following algebraic rules for isolating 'x' on one side of the equation we should <u>multiply</u> the <u>reciprocal</u> of 50 (which is 1/50) by the values on each side of the = sign.

$$\frac{1}{50} \cdot 50x = \frac{1}{50} \cdot 500$$

$$\frac{50x}{50} = \frac{500}{50} \quad x = 500 \div 50 \quad x = 10$$

The method of multiplying by the reciprocal of the 'x' value can be confusing when in fact we can <u>divide</u> each side by the value of 'x' for the same result. This 'abbreviated' method is used in industry and is also used in this book.

4. Percentage problems are the most difficult to solve, therefore, it is essential that a method be selected that works best for you and then practice with its use.

   Mastering percentage problems is a much sought after skill by employers.

   In the equation for proportion, both sides of the = sign must be equals, i.e.,

$$\frac{10}{100} = \frac{5}{50}$$
check by cross multiplying
$500 = 500$
both sides are equal proving the proportion
$\frac{10}{100} = \frac{5}{50}$ is correct.

Always check your work in this manner by substituting for x in the equation.

Direct Ratio

In the direct ratio method, we compare two numbers, as 1 and 4, by placing 1 over 4, the fraction 1/4 is formed. The top number is 1/4 the value of the bottom number, therefore, the bottom number is 4 times the value of the top number.

Caution is advised in using the direct ration method.

Example:

What percentage of unproductive land would 8 acres be of a farm with 80 acres?

Solution: (we say 8 is to 80 acres)

$$8:80 = \frac{8}{80} = \frac{1}{10} = .10 = 10\% \text{ of unproductive land}$$

In direct ratio a fraction is formed then converted to its decimal equivalent, then finally converted to a percentage.

A more difficult problem would be 3 acres of unproductive land of an 80 acre farm.

$$3:80 = \frac{3}{80} = .0375 = 3.75 \text{ or } 3\text{-}3/4\%$$

Complete the following table:

	Decimal		Percent
a.	.931	=	
b.	.19	=	
c.		=	150%
d.		=	.5%

	Fraction		Percent
e.	8/8	=	
f.	5/20	=	
g.		=	75%
h.		=	65%

Solution

a. .931, move decimal point two places to right and add
   % sign (93.1%)

b. .19, move decimal point two places to right and add
   % sign (19%). Now remove decimal point because there
   are no digits to the right.

c. 150%, place decimal point two places to the left of
   % sign and then remove % sign (1.50)

d. .5%, move decimal point two places to the left and
   remove % sign (.005)

e. 8/8 = 1 or 100%

$$8 \overline{)\overset{1.}{8.}}$$ move decimal two places
= 100%

f. 5/20 = 1/4 or 25%

$$4 \overline{)\overset{.25}{1.00}} = 25\%$$

g. 75% = 3/4

75/100 = reduce

$$\frac{75}{100} = \frac{3}{4}$$

h. 65% = $\frac{65}{100}$ or $\frac{13}{20}$

$\frac{65}{100}$ = reduce $\frac{65}{100} = \frac{13}{20}$

i. Complete the following table:

	Decimal		Percent
1.	.741	=	
2.	.29	=	
3.		=	125%
4.		=	1%

	Fraction		Percent
5.	4/4	=	
6.	4/20	=	
7.		=	25%
8.		=	55%

Find the percentage (rate) in the following problems using
all **three** methods:

Example:

1. 15 is _____% of 60?

Solution: **Business** **Method**

a. Use this formula $R = \frac{P}{B}$

b. $R = \frac{P}{B}$

c. $R = \frac{15}{60}$

$$60 \overline{)15.0} \quad .25 = 25\%$$
$$\underline{12\ 0}$$
$$3\ 00$$

d. $R = 25\%$

Solution: **Algebraic** **Method**

a. Use ratio and proportion

b. 60:100::15:x   (60 is to 100% as 15 is to x%)

c. $\frac{60}{100} = \frac{15}{x}$ set the proportion

d. 60x = 1500   cross multiply

e. $\frac{60x}{60} = \frac{1500}{60}$ divide both sides by 60

$$60 \overline{)1500} \quad 25 = 25\%$$
$$\underline{120}$$
$$300$$

f. x = 25%

Solution: **Direct** **Ratio**

a. $\frac{15}{60}$ or $\frac{1}{4}$ simplify (say 15 is to 60)

b. 15 is 25% of 60

$$4 \overline{)1.00} \quad .25 = 25\%$$
$$\underline{8}$$
$$20$$
$$\underline{20}$$

141

15 is _____% of 75

Solution: Business Method

a.

Solution: Algebraic Method

a.

Solution: Direct Ratio

a.

12 is _____% of 120

Solution: <u>Business</u> <u>Method</u>

a.

Solution: <u>Algebraic</u> <u>Method</u>

a.

Solution: <u>Direct</u> <u>Ratio</u>

a.

Find the part (of the base) in the following problems using all three methods:

Example:

15% of 60 = _____

Solution: Business Method

a. Use this formula P = RB

b. P = RB

$$\begin{array}{r} .15 \\ 60 \\ \hline 9.00 \end{array}$$

c. P = .15 x 60

d. P = 9

Solution: Algebraic Method

a. Use ratio and proportion

b. 60:100::x:15 (60 is to 100% as x is to 15%)

c. $\dfrac{60}{100} = \dfrac{x}{15}$ set the proportion

d. 100x = 900   cross multiply

e. $\dfrac{100x}{100} = \dfrac{900}{100}$ divide both sides by 100

$$\begin{array}{r} 9. \\ 100\ \overline{)900.} \\ 900 \end{array}$$

f. x = 9

Solution: Direct Ratio

a. $\dfrac{60}{100}$ or $\dfrac{6}{10}$ or $\dfrac{3}{5}$

b. $\dfrac{3}{5}$ x 15 = $\dfrac{45}{5}$ = 9

c. 9 is 15% of 60

20% or 75 = _____ .

Solution: <u>Business</u> <u>Method</u>

a.

Solution: <u>Algebraic</u> <u>Method</u>

a.

Solution: <u>Direct</u> <u>Ratio</u>

a.

10% or 120 = _____

Solution: <u>Business</u> <u>Method</u>

a.

Solution: <u>Algebraic</u> <u>Method</u>

a.

Solution: <u>Direct</u> <u>Ratio</u>

a.

Find the <u>base</u> in the following problems using all <u>three</u>
methods.

Example:

5 is 20% of _____

Solution: <u>Business</u> <u>Method</u>

a. Use this formula $B = \frac{P}{R}$

b. $B = \frac{5}{.20}$

c. B = 25

$$.20 \overline{)500.} \quad \begin{array}{r} 25. \\ \hline 500. \\ 40 \\ \hline 100 \\ 100 \end{array}$$

Solution: <u>Algebraic</u> <u>Method</u>

a. Use ratio and proportion

b. 5:20::x:100   set the ratio

c. $\frac{5}{20} = \frac{x}{100}$ set the proportion

d. 20x = 500   cross multiply

e. $\frac{20x}{20} = \frac{500}{20}$ divide both sides by 20

$$20 \overline{)500.} \quad \begin{array}{r} 25. \\ \hline 500. \\ 40 \\ \hline 100 \end{array}$$

f. x = 25

Solution: <u>Direct</u> <u>Ratio</u>

a. $\frac{5}{20} = \frac{1}{4}$

b. $\frac{1}{4} \times 100\% = \frac{100}{4}$

c. 25

147

15 is 25% of _____

Solution: <u>Business</u> <u>Method</u>

a.

Solution: <u>Algebraic</u> <u>Method</u>

a.

Solution: <u>Direct</u> <u>Ratio</u>

a.

9 is 20% of _____

Solution: <u>Business</u> <u>Method</u>

a.

Solution: <u>Algebraic</u> <u>Method</u>

a.

Solution: <u>Direct</u> <u>Ratio</u>

a.

In a welding shop, 6 circular steel discs with diameters of 4"
are stamped from steel plates 4-1/4" x 24-7/8" in size. What
is the percentage of waste?

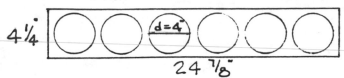

Solution:

Find the area of the discs by using this formula

$A = \pi r^2$

1. $A = \pi r^2$

2. $A = 3.14 \cdot 2^2$

3. $A = 3.14 \cdot 4 = 12.56$ sq. in.

4.   The area of 6 discs is 6 x 12.56 = 75.36 sq. in.

Find the area of the plate by using this formula: A = lw
(change fractions to decimals)

a.   A = lw

b.   A = 24.875 x 4.25 = 105.71 sq. in.

c.   Waste = 105.71 - 75.36 = 30.35 sq. in.

Find the percentage of waste using ratio and proportion

1.   105.71:100::30.35:x

2. $\dfrac{105.71}{100} = \dfrac{30.35}{x}$

3.   105.71x = 3035   cross multiply

4. $\dfrac{105.71x}{105.71} = \dfrac{3035}{105.71}$ divide both sides by 105.71

5.   x = 28.71% of waste

1. In the preceding problem, change the 6 discs diameters to 6" and the steel plate to 6-1/4" x 36-7/8" and proceed to find the percentage of waste.

   Solution.

An iron worker is placing reinforcing steel in a form for concrete. The bars are 20' long and each bar must lap the next bar 12" at each end. What percentage is lost to laps?

$$\frac{18'}{20'}$$

Solution:

Use ratio and proportion

1. 20:100::2:x

2. $\dfrac{20}{100} = \dfrac{2}{x}$

3. 20x = 200   cross multiply

4. $\dfrac{20x}{20} = \dfrac{200}{20}$ divide both sides by 20

5. x = 10% for laps

1. In the preceding problem, change the length of bars to
   40' and proceed to find the percentage lost to laps.

   Solution:

A trucker hauls gravel for a ready-mix concrete company. The bed of his truck is 7' x 20' x 4' high. By adding 12" to the height, how many more cubic yards of gravel can he haul? What percentage increase would this be?

Solution:

Find the cubic yards that can be carried by using volume formula V = lwh

1. $V = lwh$

2. $V = 20 \cdot 7 \cdot 4$

3. $V = 560$ cubic feet $\div 27 = 20.74$ cubic yards

Find the percentage of increase

a. $V = lwh$

b. $V = 20 \cdot 7 \cdot 5$ (5 represents 5' side height)

c. $V = 700$ cubic feet $\div 27 = 25.92$ cubic yards

d. Use ratio and proportion

e. $560:100::140:x$ (700 ft.^{3} - 560 ft.^{3} = 140 ft.^{3} increase)

f. $\dfrac{560}{100} = \dfrac{140}{x}$

g. $560x = 14000$ cross multiply

h. $\dfrac{560x}{560} = \dfrac{14000}{560}$ divide both sides by 560

i. $x = 25\%$ increase

1. In the preceding problem, change the bed size to 7' x
   24' x 5' and increase to 6' high. Proceed with solution.

   Solution:

A farmer's corn storage silo is 8' in diameter and 20' tall.
How many cubic feet of newly shelled corn can be stored?
Corn shrinks about 7% after drying. How many cubic feet of
corn are lost by drying?

Solution:

Find the volume of the silo

1. Use volume formula $V = \pi r^2 h$

2. $V - \pi r^2 h$

3. $V = 3.14 \cdot 4^2 \cdot 20$

4. $V = 3.14 \cdot 16 \cdot 20$

5. $V = 1004.8$ cubic feet (round to 1005)

Find the percentage of waste in cubic feet

a. Use ratio and proportion

b. $1005:100::x:7$

c. $\dfrac{1005}{100} = \dfrac{x}{7}$

d. $100x = 7035$ cross multiply

e. $\dfrac{100x}{100} = \dfrac{7035}{100}$ divide both sides by 100

f. $x = 70.35$ cubic feet lost to drying

A tank manufacturer allows for expansion by providing a 3% increase in rated capacity. A tank with a diameter of 4' and 8' long holds how many gallons? Allowing for expansion how many additional gallons are allowed?

Solution:

1. Find the volume of the tank with this formula:

$$V = \pi r^2 h$$

2. $V = \pi r^2 h$

3. $V = 3.14 \cdot 2^2 \cdot 8$

4. $V = 3.14 \cdot 4 \cdot 8$

5. $V = 100.48$ cubic feet

6. $V = \dfrac{100.48}{7.48} = 13.4$ gallons (1 cubic foot = 7.48 gallons)

7. Provide 3% for expansion

8. $13.4 \times .03 = .4$ of one gallon

9. $13.4 + .4 = 13.8$ gallons

or

Use ratio and proportion to find 3% for expansion

a. $13.4:100::x:3$ (13.4 is to 100% as x is to 3%)

b. $\dfrac{13.4}{100} = \dfrac{x}{3}$

c. $100x = 40.2$ cross multiply

d. $\dfrac{100x}{100} = \dfrac{40.2}{100}$ divide both sides by 100

e. $x = .402$ or .4 of a gallon

f. Rated capacity would be $13.4 + .4 = 13.8$ gallons

Gasoline stored in uninsulated above ground tanks lose approximately 3% to evaporation in a given amount of time. How many gallons would a 40' diameter by 30' high tank hold? How many gallons are lost to evaporation?

Solution:

Find the volume of the tank using this formula:

$V = \pi r^2 h$

1. $V = \pi r^2 h$

2. $V = 3.14 \cdot 20^2 \cdot 30$

3. $V = 3.14 \cdot 400 \cdot 30$

4. $V = 37680$ cubic feet

5. Gallons $= \dfrac{37680}{7.48} = 5037.4$

6. There are 5037.4 (round to 5037 gallons)

Find the amount lost to evaporation

a. Use ratio and proportion

b. $5037:100::x:3$

c. $\dfrac{5037}{100} = \dfrac{x}{3}$

d. $100x = 15111$ cross multiply

e. $\dfrac{100x}{100} = \dfrac{15111}{100}$ divide both sides by 100

f. $x = 151.11$ (round to 151 gallons lost to evaporation)

158

In the preceding problem, change the size of the tank to 10' diameter and 24' tall. Proceed with the solution.

Solution:

A welder received a base rate of 12.00 per hour. He received a production bonus of 2.00 per hour for his work this week. What percent of his base did this represent?

Solution:

Use ratio and proportion.

1. 12.00:100::2.00:x   set the ratio

2. $\dfrac{12.00}{100} = \dfrac{2.00}{x}$ set the proportion

3. 12.00x = 200   cross multiply

4. $\dfrac{12.00x}{12.00} = \dfrac{200}{12.00}$ divide both sides by 12.00

5. $x = \dfrac{200}{12.00}$

6. x = 16.66% rate of production bonus

1. In the preceding problem, change the base rate to 13.50 per hour and the production bonus to 2.70 per hour. Proceed with the solution.

    Solution: (Use algebraic method)

2. Solution: (Use business method)

    Use the formula $R = \dfrac{P}{B}$

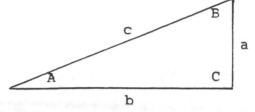

UNIT 14

Right Triangles:

Trigonometry is a study of angles and their functions.

Right-angle trigonometry is used to find unknown points of a right triangle. These can be sides (including the hypotenuse) or angles. Triangles that do not contain a right angle (90°) are called oblique triangles. This unit presents techniques to solve both right triangles and oblique triangles.

Be patient with yourself when working the problems in this unit as time will be needed to thoroughly digest the solving processes. Use your calculator when working these problems.

You may also use the Trig table at the end of this unit to find the _nearest_ degree and how to determine minutes of a degree.

Function		Given	To Find A	To Find B
SIN ∠	=	Side Opposite / Hypotenuse	SIN A = $\dfrac{a}{c}$	SIN B = $\dfrac{b}{c}$
COS ∠	=	Side Adjacent / Hypotenuse	COS A = $\dfrac{b}{c}$	COS B = $\dfrac{a}{c}$
TAN ∠	=	Side Opposite / Side Adjacent	TAN A = $\dfrac{a}{b}$	TAN B = $\dfrac{b}{a}$
COT ∠	=	Side Adjacent / Side Opposite	COT A = $\dfrac{b}{a}$	COT B = $\dfrac{a}{b}$
SEC ∠	=	Hypotenuse / Side Adjacent	SEC A = $\dfrac{c}{a}$	SEC B = $\dfrac{c}{a}$
CSC ∠	=	Hypotenuse / Side Opposite	CSC A = $\dfrac{c}{a}$	CSC B = $\dfrac{c}{b}$

Given	To Find Side a	To Find Side b	To Find Side c
A & side c	SIN A·c = a		
A & side c		COS A·c = b	
A & side a			$\dfrac{a}{SIN\ A} = c$

162

Example:

Find $\angle$A if a = 8 and c = 14

Use the formula $SIN\ A = \dfrac{a}{c}$

$\qquad SIN\ A = \dfrac{a}{c}$

$\qquad SIN\ A = \dfrac{8}{14}$

$\qquad SIN\ A = .57142$

$\qquad A = 35^{\circ}$ (from Trig table)

Find $\angle$B if a = 8 and c = 14

Use the formula $COS\ B = \dfrac{a}{c}$

$\qquad COS\ B = \dfrac{a}{c}$

$\qquad COS\ B = \dfrac{8}{14}$

$\qquad COS\ B = .57142$

$\qquad B = 55^{\circ}$

Find side a if $A = 35^{\circ}$ and c = 14

Use the formula $SIN\ A \cdot c = a$

$\qquad SIN\ A \cdot c$

$\qquad .57142 \cdot 14 = a$

$\qquad a = 7.999$ (round to 8)

Find side b if A = 35° and c = 14

Use the formula COS A · c = b

  COS A · c = b

  .82065 · 14 = 11.489 (round to 11.5)

Find side c if A = 35° and a = 8

Use the formula $\dfrac{a}{SIN\ A}$ = c

  $\dfrac{a}{SIN\ A}$ = c

  $\dfrac{8}{.57142}$ = c

  c = 14.0

NOTE: To verify that a = 8, b = 11.5 and c = 14, use the Pythagorean formula for right triangles.

  $c^2 = a^2 + b^2$

  $c^2 = 64 + 132.25$

  $c^2 = 196.25$

  $c = \sqrt{196.25}$

  c = 14.008 (round to 14)

  $a^2 = c^2 - b^2$ $b^2 = c^2 - a^2$

  $a^2 = 196.25 - 132.25$ $b^2 = 196.25 - 64$

  $a^2 = 64$ $b^2 = 132.25$

              $b = \sqrt{132.25}$

  $a = \sqrt{64}$ b = 11.5

  a = 8

164

To verify that $A = 35°$ and $B = 55°$ use $\angle\underline{s}$ of a triangle $= 180°$

$c = 90°$ (the right angle)

Therefore, $180° - 90° = 90°$

$A + B = 90°$

Therefore, $90° - A = B$ (or $90° - B = A$)

$90° - 35° = 55°$

or

$90° - 55° = 35°$

Use
$$\begin{array}{r} 90° \\ - \underline{35°} \quad A \\ 55° \quad B \end{array}$$

or

$$\begin{array}{r} 90° \\ - \underline{55°} \quad B \\ 35° \quad A \end{array}$$

1. An entrance ramp to the interstate is 700' long and rises 30'. What is the degree of rise?

Solution:

Given a = 30', c = 700'

Use the formula $SIN \ A = \dfrac{a}{c}$

$SIN \ A = \dfrac{a}{c}$

$SIN \ A = \dfrac{30}{700}$

$SIN \ A = .04285$

$A = 2°$ (from Trig table)

The degree of rise is 2° in 700'

2. Find $\angle B$ if the entrance ramp is 1500' long and rises 70'.

Solution: Use $COS \ B = \dfrac{a}{c}$

1. A drafter is preparing a scale drawing (1" = 500') of a
    section of the interstate highway that is 1-1/2 miles
    long and rises 550'. He needs to know the degree of
    rise; also when drawn at the scale 1" = 500', how long
    would the line be?

    Solution:

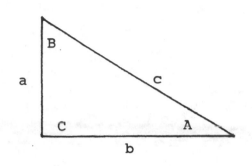

1. Find side c if a = 9 and A = 60°. Use this formula,

   $$c = \frac{a}{SIN\ A}$$

   Solution:

2. Find side a if A = 30° and c = 3. Use this formula,

   a = SIN A · c

   Solution:

1. Find side b if c = 12 and A = 55°. Use this formula,

   b = COS A · c

   Solution:

2. Find angle A if side a = 8 and side c = 12. Use this formula.

   SIN A = $\frac{a}{c}$

   Solution:

1. Find angle B if side b = 7 and side a = 10. Use this
   formula.

   TAN B = $\dfrac{b}{a}$

   Solution:

2. Find side c if side a = 5 and side b = 3. Use this
   formula.

   $c = \sqrt{a^2 + b^2}$

   Solution:

Oblique Triangles

LAW OF SINES: in any triangle the sides are in the same ratio
                as the sines of the angles opposite them.

> SOLUTION for oblique triangles when two angles and a side
> are given.

$$\frac{b}{\text{Sin } B} = \frac{c}{\text{Sin } C}$$

$$\frac{b}{\text{Sin } B} = \frac{d}{\text{Sin } D}$$

$$\frac{c}{\text{Sin } C} = \frac{d}{\text{Sin } D}$$

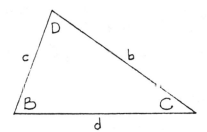

> SOLUTION for oblique triangles when two sides and an angle
> opposite one of them is given.

(Equations are the same as above except reverse the ratio;
see example page 163.)

LAW OF COSINES: in any triangle the square of any side equals
                  the sum of the squares of the other two sides
                  diminished by twice the product of these two
                  sides multiplied by the cosine of the included
                  angle.

> SOLUTION for oblique triangles when two sides and the in-
> cluded angle are given.

$$b^2 = c^2 + d^2 - 2cd \text{ COS } B$$

$$c^2 = b^2 + d^2 - 2bd \text{ COS } C$$

$$d^2 = b^2 + c^2 - 2bc \text{ COS } D$$

NOTE:

After finding the third side of the triangle, use the law
of sines to find the other angles.

The cosine of an angle between 90° and 180° is a negative
number.

SOLUTION for oblique triangles when all three sides are
given.

$$\text{COS } B = \frac{c^2 + d^2 - b^2}{2cd}$$

$$\cos C = \frac{b^2 + d^2 - c^2}{2bd}$$

$$\cos D = \frac{b^2 + c^2 - d^2}{2bc}$$

Example:

Given: B = 80°

C = 25°

b = 14

Find side c

Use the equation $\dfrac{b}{\text{SIN } B} = \dfrac{c}{\text{SIN } C}$

$$\frac{b}{\text{SIN } B} = \frac{c}{\text{SIN } C}$$

$$\frac{14}{\text{SIN } 80°} = \frac{c}{\text{SIN } 25°}$$

$$\frac{14 \cdot \text{SIN } 25°}{\text{SIN } 80°} = c \quad \text{(cross multiply)}$$

$$\frac{14 \cdot .42262}{.98481} = c$$

$$\frac{.98481c}{.98481} = \frac{5.91668}{.98481} \quad \text{divide both sides by .98481}$$

$$\frac{5.91668}{.98481} = 6.007 \quad \text{(round to 6)}$$

c = 6

Example:

Given $D = 75°$
               $C = 25°$
               $c = 6$

Find side d

Use the equation $\dfrac{c}{\text{SIN } C} = \dfrac{d}{\text{SIN } D}$

$$\frac{c}{\text{SIN } C} = \frac{d}{\text{SIN } D}$$

$$\frac{6}{\text{SIN } 25°} = \frac{d}{\text{SIN } 75°}$$

$$\frac{6 \cdot .96593}{.42262} = d$$

$$\frac{5.79558}{.42262} = 13.713 \text{ (round to 13.7)}$$

$d = 13.7$

A pattern is being made but the patternmaker needs to find B.

Given: $c = 12$
                $b = \ \ 9$
                $C = 40°$

Use this equation and reverse the ratio

$$\frac{b}{SIN\ B} = \frac{c}{SIN\ C}$$

$$\frac{SIN\ B}{b} = \frac{SIN\ C}{c}$$

$$\frac{SIN\ B}{b} = \frac{SIN\ 40°}{c}$$

$$\frac{9 \cdot .64279}{12} = SIN\ B$$

$$\frac{5.78511}{12} = SIN\ B$$

$$.48209 = SINB$$

$$B = 28° \text{ (from Trig. table)}$$

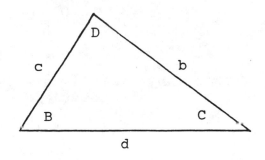

1. Find angle B if A = 70° and C = 65°.   Use this formula,
    B = 180° - (A + C)

    Solution:

2. Find side c if side d = 10, angle C = 75° and angle D = 65°.   Use this formula,

    c = $\dfrac{\text{SIN C} \cdot \text{d}}{\text{SIN D}}$

    Solution:

1. Find TAN A if side d = 9, angle B = 80° and side c = 7.   Use this formula,

    $$\text{TAN A} = \frac{d \cdot \text{SIN B}}{c - (d \cdot \text{COS B})}$$

    Solution:

2. Find side b if side d = 11, Angle D = 62° and angle B = 75°.   Use this formula,

    $$b = \frac{d \cdot \text{SIN B}}{\text{SIN D}}$$

    Solution:

TABLE OF NATURAL TRIGONOMETRIC FUNCTIONS

This is a commonly used table to locate degrees and functions. Larger tables provide for more accurate readings, locating degrees and each minute in a degree. However, the following example will show how to find minutes of a degree using <u>this</u> table. (Refer to page 163 for the example.)

As shown, the Sine of A = .57142, round to 4 places, .5714 and find under the Sine column. The number is between 34° and 35°. 34° = Sine of .5592 and 35° = Sine of .5736; subtract these two numbers to get 144. The number .5714 is 122 points greater than 34°. Set up ratio and proportion as follows: (144 is to 60 minutes as 122 is to X minutes; 144:60::122:X) One degree = 60 minutes.

$$\frac{144}{60} = \frac{122}{X}$$

144 X = 7320 (Cross multiply)

$$\frac{144 \, X}{144} = \frac{7320}{144} \quad X = 50.8 \text{ or } 51' \quad \text{(Rounded)}$$

Therefore the Sine of .57142 (Rounded to .5714) = 34°51'. This number was rounded to 35° in the problem on page 163.

Angle	Sine	Cosine	Tangent	Angle	Sine	Cosine	Tangent	Angle	Sine	Cosine	Tangent
1°	.0175	.9998	.0175	31°	.5150	.8572	.6009	61°	.8746	.4848	1.8040
2°	.0349	.9994	.0349	32°	.5299	.8480	.6249	62°	.8829	.4695	1.8807
3°	.0523	.9986	.0524	33°	.5446	.8387	.6494	63°	.8910	.4540	1.9626
4°	.0698	.9976	.0699	34°	.5592	.8290	.6745	64°	.8988	.4384	2.0503
5°	.0872	.9962	.0875	35°	.5736	.8192	.7002	65°	.9063	.4226	2.1445
6°	.1045	.9945	.1051	36°	.5878	.8090	.7265	66°	.9135	.4067	2.2460
7°	.1219	.9925	.1228	37°	.6018	.7986	.7536	67°	.9205	.3907	2.3559
8°	.1392	.9903	.1405	38°	.6157	.7880	.7813	68°	.9272	.3746	2.4751
9°	.1564	.9877	.1584	39°	.6293	.7771	.8098	69°	.9336	.3584	2.6051
10°	.1736	.9848	.1763	40°	.6428	.7660	.8391	70°	.9397	.3420	2.7475
11°	.1908	.9816	.1944	41°	.6561	.7547	.8693	71°	.9455	.3256	2.9042
12°	.2079	.9781	.2126	42°	.6691	.7431	.9004	72°	.9511	.3090	3.0777
13°	.2250	.9744	.2309	43°	.6820	.7314	.9325	73°	.9563	.2924	3.2709
14°	.2419	.9703	.2493	44°	.6947	.7193	.9657	74°	.9613	.2766	3.4874
15°	.2588	.9659	.2679	45°	.7071	.7071	1.0000	75°	.9659	.2588	3.7321
16°	.2756	.9613	.2867	46°	.7193	.6947	1.0355	76°	.9703	.2419	4.0108
17°	.2924	.9563	.3057	47°	.7314	.6820	1.0724	77°	.9744	.2250	4.3315
18°	.3090	.9511	.3249	48°	.7431	.6691	1.1106	78°	.9781	.2079	4.7046
19°	.3256	.9455	.3443	49°	.7547	.6561	1.1504	79°	.9816	.1908	5.1446
20°	.3420	.9397	.3640	50°	.7660	.6428	1.1918	80°	.9848	.1736	5.6713
21°	.3584	.9336	.3839	51°	.7771	.6293	1.2349	81°	.9877	.1564	6.3138
22°	.3746	.9272	.4040	52°	.7880	.6157	1.2799	82°	.9903	.1392	7.1154
23°	.3907	.9205	.4245	53°	.7986	.6018	1.3270	83°	.9925	.1219	8.1443
24°	.4067	.9135	.4452	54°	.8090	.5878	1.3764	84°	.9945	.1045	9.5144
25°	.4226	.9063	.4663	55°	.8192	.5736	1.4281	85°	.9962	.0872	11.4301
26°	.4384	.8988	.4877	56°	.8290	.5592	1.4826	86°	.9976	.0698	14.3007
27°	.4540	.8910	.5095	57°	.8387	.5446	1.5399	87°	.9986	.0523	19.0811
28°	.4695	.8829	.5317	58°	.8480	.5299	1.6003	88°	.9994	.0349	28.6363
29°	.4848	.8746	.5543	59°	.8572	.5150	1.6643	89°	.9998	.0175	57.2900
30°	.5000	.8660	.5774	60°	.8660	.5000	1.7321	90°	1.0000	.0000	

APPENDIX A

WEIGHTS AND MEASURES

Volume measures and equivalents:

1728 cubic inches	=	1 cubic foot (cu. ft.)
27 cubic feet	=	1 cubic yard (cu. yd.)
2 pints	=	1 quart (qt.)
4 quarts	=	1 gallon (gal.)
1 cu. ft.	=	7.48 gal.
231 cu. in.	=	1 gal.
31-1/2 gal.	=	1 barrel (bbl.)

1 gallon of water weighs approx. 8-1/3 lbs. (8.333 lbs.)

1 cu. ft. of water weighs 62-1/2 lbs. (62.5 lbs.)

Table of specific gravities:

water	=	1.000
sulphuric acid	=	1.835
alcohol	=	.816
gasoline	=	.728

Length:

12 inches (in.)	=	1 foot (ft.)
3 feet	=	1 yard (yd.)
1 mile (mi.)	=	5280 feet (ft.)
1 mile (mi.)	=	1760 yards (yds.)

Area:

144 square inches	=	1 square foot (sq. ft.)
9 square feet	=	1 square yard (sq. yd.)
43,560 square feet	=	1 acre (a.)
640 acres	=	1 square mile (sq. mi.)

TABLE OF SQUARES AND SQUARE ROOTS

n	n^2	$\sqrt{n}$	n	n^2	$\sqrt{n}$
1	1	1.000	51	2601	7.141
2	4	1.414	52	2704	7.211
3	9	1.732	53	2809	7.280
4	16	2.000	54	2916	7.348
5	25	2.236	55	3025	7.416
6	36	2.449	56	3136	7.483
7	49	2.646	57	3249	7.550
8	64	2.828	58	3364	7.616
9	81	3.000	59	3481	7.681
10	100	3.162	60	3600	7.746
11	121	3.317	61	3721	7.810
12	144	3.464	62	3844	7.874
13	169	3.606	63	3969	7.937
14	196	3.742	64	4096	8.000
15	225	3.873	65	4225	8.062
16	256	4.000	66	4356	8.124
17	289	4.123	67	4489	8.185
18	324	4.243	68	4624	8.246
19	361	4.359	69	4761	8.307
20	400	4.472	70	4900	8.367
21	441	4.583	71	5041	8.426
22	484	4.690	72	5184	8.485
23	529	4.796	73	5329	8.544
24	576	4.899	74	5476	8.602
25	625	5.000	75	5625	8.660
26	676	5.099	76	5776	8.718
27	729	5.196	77	5929	8.775
28	784	5.292	78	6084	8.832
29	841	5.385	79	6241	8.888
30	900	5.477	80	6400	8.944
31	961	5.568	81	6561	9.000
32	1024	5.657	82	6724	9.055
33	1089	5.745	83	6889	9.110
34	1156	5.831	84	7056	9.165
35	1225	5.916	85	7225	9.220
36	1296	6.000	86	7396	9.274
37	1369	6.083	87	7569	9.327
38	1444	6.164	88	7744	9.381
39	1521	6.245	89	7921	9.434
40	1600	6.325	90	8100	9.487
41	1681	6.403	91	8281	9.539
42	1764	6.481	92	8464	9.592
43	1849	6.557	93	8649	9.644
44	1936	6.633	94	8836	9.695
45	2025	6.708	95	9025	9.747
46	2116	6.782	96	9216	9.798
47	2209	6.856	97	9409	9.849
48	2304	6.928	98	9604	9.899
49	2401	7.000	99	9801	9.950
50	2500	7.071	100	10000	10.000

FINDING THE SQUARE ROOT OF A NUMBER

Example - $\sqrt{8475}$

```
              9   2.  0   5   9
        9  √84'75!00'00'00
           -81
      18Ø2/  3 75
            -3 64
      1840  / 11 00
               -  0
   1840Ø5/ 11 00 00
           -9 20 25
   184100 / 1 79 75 00
           -1 65 69 00
                14 06 00
```

1. Divide the number into sets of two digits (if the number has an odd number of digits, the first digit is left by itself)

2. Figure the closest square root of the number in the first set, write that number above the <u>first</u> <u>set</u> and to the left of the radical. Multiply those numbers together and sub-tract the product from the first set.

3. Bring down the next set (the next <u>two</u> digits). Double your answer (ignore any decimal) and add a zero to the right. Write this number to the left of the remainder.

4. Determine about how many times the remainder can be <u>divided</u> by the number to its left. Write this number above the next set and change the last zero in the divisor to the same number.

5. Multiply the new divisor by the last digit of the answer and subtract the product from the remainder.

6. Continue repeating steps 3, 4 and 5 until finished.

The method outlined above has been around for centurys and has been a source of irritation and confusion for those working with square roots. We are fortunate now to have calculators that can work square roots with the tap of a key.

However , for those who want to review the "old" method, this page is for you. Enjoy!